The Chemists' War
1914–1918

The Chemists' War
1914–1918

Michael Freemantle
Kempshott, Basingstoke, UK
Email: michael.freemantle@btconnect.com

THE QUEEN'S AWARDS
FOR ENTERPRISE:
INTERNATIONAL TRADE
2013

Print ISBN: 978-1-84973-989-4

A catalogue record for this book is available from the British Library

Published by The Royal Society of Chemistry,
Thomas Graham House, Science Park, Milton Road,
Cambridge CB4 0WF, UK

Registered Charity Number 207890

Visit our website at www.rsc.org/books

Printed and bound by CPI Group (UK) Ltd, Croydon, CR0 4YY

Preface

The chemistry of the First World War is a field of study where two great continents of knowledge collide: namely chemistry and history. The chapters that follow are based on my personal albeit limited exploration of this vast and somewhat uncharted territory. They outline what I consider to be some of the more salient and fascinating features of the territory.

My exploration relied heavily on English language sources of information and accounts of the war. I found the journal archives of the Royal Society, the Royal Society of Chemistry, the American Chemical Society, and the British Medical Association both helpful and revealing. Much of the information extracted from these and other sources relates to the Western Front. Less information is available about the chemistry of the war in some other theatres of engagement, for example East Africa.

I have little doubt that a wealth of untapped information about the chemists and chemistry of the war can be found in archives and other sources in Athens, Berlin, Budapest, Istanbul, Moscow, Tokyo, and elsewhere. A full treatment of "the chemists' war," as World War I is sometimes known, would therefore probably require a team of dedicated researchers who are not only familiar with chemistry and the history of the war but also proficient in the languages of the war's belligerent nations. Such a treatment is inevitably beyond the scope of this book.

The Chemists' War, 1914–1918
By Michael Freemantle
© Freemantle, 2015
Published by the Royal Society of Chemistry, www.rsc.org

The book is written for the general reader and the many scientists and historians interested in the Great War of 1914–1918. It was not my aim to write a book that could be read from cover to cover but rather one for the reader to dip into. Each chapter is intended to be self-contained and can be read independently of the other chapters. As a consequence, there is a certain amount of repetition, although I have tried to keep this to a minimum. For example, I have devoted a whole chapter to Fritz Haber, the most prominent chemist of the war and possibly the most important chemist of the 20th century. Even so, his name necessarily crops up several times elsewhere in the book. Similarly, the book has a chapter on mustard gas, a topic which also appears in some other chapters.

Numerous people contributed to this book in one way or another. Many—the authors of the reports, articles, and personal accounts published in journals and books during or just after the war—are now dead. For the chapter on women's contributions to the war, I relied heavily on the work and publications of Geoffrey Rayner-Canham, chemistry professor at Memorial University of Newfoundland, Canada, and his co-researcher Marelene Rayner-Canham. I am grateful to: Philip Hoare, author and self-professed whale obsessive, for sparking my interest in the connection between whaling and the chemistry of the war.

For their assistance with photos and permissions to publish them, I would like to thank: Anne Barrett, College Archivist, Imperial College, London; David Bell, Commercial Sales & Licensing Manager, Imperial War Museums, London; Naama Chomski-Pesso of the Weizmann Institute of Science, Rehovot; Thierry Gourlin, Président, Musée Somme 1916; Hillary Kativa, Archivist at the Othmer Library of Chemical History, Chemical Heritage Foundation, Philadelphia; Anne Thomson, College Archivist, Newnham College, Cambridge; Susanne Uebele, Archivist, Max-Planck-Gesellschaft, Berlin-Dahlem; and Carol Walker, Director, The Somme Association.

At the Royal Society of Chemistry (RSC), Burlington House, London, David Allen, Library Collections Coordinator, and Kate Bennett, Library Operations Specialist, provided essential guidance for the chapter on the two RSC war memorials. I would also like to thank the following staff at the RSC in Cambridge for their invaluable help: Janet Freshwater, Senior Commissioning Editor;

Sylvia Pegg, Senior Production Controller, Books; and Helen Prasad, Books Development Administrator.

Finally, I would like to acknowledge the considerable support of my wife, Mary, and dedicate this book to our youngest grandchild, Lucien Dominic.

Michael Freemantle
Basingstoke

About the Author

Dr Michael Freemantle is a science writer and a Fellow of the Royal Society of Chemistry. After a post-doctoral research fellowship at Oxford University (1967–1969), he worked in the chemical industry for two years. From 1971 to 1985, he taught chemistry at various levels both in the UK and abroad. In 1985, he was appointed Information Officer for IUPAC (International Union of Pure and Applied Chemistry). His duties included editing the IUPAC news magazine *Chemistry International*. From 1994 to 2007 he was European Science Editor/Senior Correspondent for *Chemical & Engineering News*—the weekly news magazine of the American Chemical Society. He was then appointed Science Writer in Residence, a part-time post, at Queen's University Belfast and Queens University Ionic Liquid Laboratories for three years until 2010.

Freemantle has written a number of books and numerous news reports and articles on chemistry, the history of chemistry, and related topics. He is the author of: *An Introduction to Ionic Liquids*, RSC Publishing, 2009; and *Gas! GAS! Quick, boys! How Chemistry Changed the First World War*, The History Press, 2012.

Contents

The Chemists' War, 1914–1918
By Michael Freemantle
© Freemantle, 2015
Published by the Royal Society of Chemistry, www.rsc.org

CHAPTER 1

Much More than Chemical Warfare

THE CHEMISTRY OF THE GREAT WAR

When I tell people that I'm fascinated by the Great War, they typically respond with a comment such as: "Me too! I was so fond of my grandfather who fought in the war. He lost a leg in the trenches when he was nineteen." They then enquire if I've seen *War Horse*, a play adapted from a novel by Michael Morpugo and subsequently made into a film, or read the novel *Birdsong* by Sebastian Faulks which has also been dramatized. Both are set on the Western Front.

If I mention that I'm writing a book on the chemistry of the First World War, they are likely to relate to me an account of how a great uncle or some other person they knew was gassed in the war. They understandably associate the chemistry of the war with chemical warfare, that is the deployment of chemical weapons, or poison gases as they are often called, against the enemy combined with the use of gas masks and other protective anti-gas measures for defence.

Yet, as I explain to them, chemical warfare was just one component, albeit an important component, of the chemistry of the war. But it wasn't the most important. The opposing armies spent much of their time firing other lethal chemicals at one

The Chemists' War, 1914–1918
By Michael Freemantle
© Freemantle, 2015
Published by the Royal Society of Chemistry, www.rsc.org

Figure 1.1 A First World War trench mortar and high explosive shell on display at the Ulster Tower Memorial, Thiepval, northern France (author's own, 2010).

another, most notably powerful high explosives such as trinitrotoluene (TNT). And to fire shells or mortar bombs filled with these explosive chemicals, other chemicals were also needed, the most important of which were propellants like cordite.

ONE HUNDRED SHELLS A MINUTE

World War I was a conflict between the Allied Powers and the Central Powers that took place from late July 1914 to November 1918. The Allied Powers consisted of Britain, France, Japan, Russia and Serbia. Italy joined in 1915, Portugal and Romania in 1916, and Greece and the United States in 1917. The Central Powers were Germany, the Austro-Hungarian Empire and Ottoman Turkey, with Bulgaria joining in 1915.

Millions upon millions of artillery shells filled with high explosives were fired in the war. At times, the shelling was unimaginably intense. In preparation for the Battle of the Verdun,

Figure 1.2 Variety of shells on display at The Somme Trench Museum, Albert, France (author's own, 2010, Musée Somme, 1916).

for instance, the Germans stockpiled around three million shells. The battle, the longest of the war, was fought at Verdun-sur-Meuse, a city in north-eastern France, from 21 February to 18 December 1916. In the first hour of the attack alone, the German army fired a staggering 100 000 shells at the French who were defending the city. Over the course of the 10 month battle, the two armies bombarded each other with some 23 million shells. That works out at about 100 shells per minute, day and night. Each shell was filled with chemicals of one sort or another and propelled out of the muzzles of gun barrels by gases at high pressure generated by firing propellants. The battle resulted in almost one million French and German casualties, over half of whom died.

The shelling was similarly intense the following year during the Battle of Messines that took place from 7–14 June along Messines Ridge, south of Ypres, Belgium. Before the battle, British artillery bombarded German defences on the ridge with 3.5 million shells over a ten day period. That equates to three shells per second. Furthermore, army tunnelling engineers from Australia, Britain, Canada and New Zealand laid 22 mines along a 10 mile front prior to the battle. They contained 450 tons of ammonal, an explosive containing ammonium nitrate and aluminium. The Germans detonated one of the mines in a countermining operation. Of the remaining mines, 19 exploded killing or burying alive about 10 000 German troops and wrecking German positions as well as the town of Messines.

Between 4 August 1914, when Britain declared war on Germany, and the Armistice on 11 November 1918, over nine million men in the forces of the Allies and Central Powers were killed in battle (Figure 1.3). Artillery shelling accounted for an estimated 58% of the battlefield deaths, machine-gun and rifle fire for 39%, and poison gas less than 3%. It should be noted in passing, however, that these percentages along with many other statistics of the war are no more than rough estimates. Death tolls and causes of death varied from year to year and from front to front, and records, when kept, were not always accurate and even if accurate were sometimes lost or destroyed.

Nevertheless, it is indisputable that death on the battlefields of the war occurred on an industrial scale and that this horrendous slaughter relied on the industrial-scale production of explosives, poison gases, and other chemicals. Chemical plants throughout Europe worked frantically to manufacture these deadly chemicals in sufficient quantities to meet the rapacious demands of munitions factories and the armed forces.

In North Wales, for instance, a plant at Queensferry near Chester produced 250 tons of nitrocellulose, a high explosive also known as guncotton, and 500 tons of TNT each week at the height of the war. There were many other such plants in Britain. Some plants specialised in filling shells with explosives. The National Filling Factory No.6 at Chilwell in Nottinghamshire, England, is just one example. The factory filled more than 19 million shells, 25 000 sea mines, and 2500 bombs with explosives during the war.[1]

CHEMICAL WARFARE

The Allies and Central Powers employed a wide range of chemical warfare agents in the war. Chlorine, a gas that caused troops

Figure 1.3 Over 1200 Commonwealth servicemen of the First World War are buried or commemorated at the Connaught Cemetery, Thiepval, France (author's own, 2010).

to choke when inhaled in high concentrations, was first used by the Germans on the Western Front on 22 April 1915. Under the supervision of German chemist Fritz Haber (1868–1934), gas troops unleashed almost 170 tons of the gas from cylinders against the Allies at Langemarck, a village north of Ypres, killing as many as 6000 French, Moroccan, and Algerian troops. It was the first time in history that a weapon of mass destruction had been used.

On 25 September that year, the British retaliated with their own gas attacks on the first day of the Battle of Loos in northern France. They released some 140 tons of chlorine from just under 6000 cylinders and at the same time used mortars to fire around 10 000 smoke bombs filled with phosphorus. Although some of the gas blew back over British trenches, the attack caused panic and confusion among the German troops and the British were able to break through their lines and capture the town of Loos on the same day.

With the development of masks to protect against chlorine gas, other poison gases were soon introduced: phosgene in December 1915, diphosgene in May 1916, and chloropicrin in August 1916. The Germans introduced mustard gas in July 1917 when they bombarded the British frontlines near Ypres with an estimated 50 000 mustard gas shells.

Cloud gas operations continued throughout the war using either chlorine or mixtures of chlorine with other toxic gases. In these operations, the gases were released from cylinders. Millions of artillery shells containing tear gases and lethal gases such as phosgene and diphosgene were also fired by all sides during the war. In the 1917 Messines offensive, for example, the British fired 75 000 shells containing a total of more than 160 tons of chloropicrin against the Germans.

Altogether, the Germans, French, and British manufactured around 68 000 tons, 37 000 tons, and 26 000 tons of these so-called "battle gases", respectively, during the war. Austria, Italy, Russia and the United States together manufactured just over 19 000 tons. That all adds up to a total of 150 000 tons, of which 25 000 tons were "left on hand unused" at the end of the war, according to one source.[2]

MANY OTHER CHEMICALS

The chemistry of the war didn't stop with the manufacture of chemical warfare agents and the propellants and explosives needed for weapons and ammunition. Star or illumination shells, signal rockets, and Very flares all contained pyrotechnic compositions similar to those used in fireworks today. When ignited, compositions containing barium chlorate, copper carbonate, strontium carbonate, and magnesium powder produced green, blue, red, and white illuminations, respectively. Very flares were launched by firing cartridges from flare guns known as Very pistols. Each cartridge contained a paper cylinder containing the pyrotechnic mixture, a propellant charge of gunpowder, and a percussion cap that ignited the propellant when the pistol was fired. The flares were named after their inventor, American naval officer Edward W. Very (1847–1910).

Other chemical materials were also essential for the war effort. Iron, steel, bronze, brass, gunmetal, copper, and aluminium, all relied on chemical processes for their production. The metals and alloys were needed for the manufacture of guns, ammunition, armour, military aircraft, battleships, tanks, and submarines, and the railways that carried troops, horses, and supplies to the front and sick and wounded troops back from the front.

We should not forget dyes. Every single soldier, sailor, airman, medical officer, and nurse in the war wore a uniform of one sort or another. British, American, and Russian soldiers, for example, wore khaki uniforms. According to some estimates, around 43 million men in the Allied forces and 25 million men in the forces of the Central Powers fought in the war. That adds up to 68 million military uniforms—all coloured by synthetic or naturally-occurring chemical dyes or mineral pigments.

Khaki and other drab colours, such as the field grey worn by the German army, provided camouflage and therefore helped to protect troops behind the lines, in the trenches, and on the battlefields. Chemical materials were also exploited for a variety

Figure 1.4 Soldiers in a firing trench in the Dardanelles. Every single uniform was coloured by a synthetic or naturally-occurring chemical dye or mineral pigment (The RAMC Collection in the care of the Wellcome Library).

of other protective roles, for instance in gas masks. British gas masks introduced in 1916 incorporated canisters containing a variety of chemicals. One of them was a highly porous form of carbon known as activated charcoal. The powder trapped poison gases inside its pores while allowing air to get through.

Some chemicals were used for both destructive and protective roles. Chlorine, for instance, was used not only as a war gas but also to save lives. The element or chemicals such as bleaching powder that released the element were used to purify drinking water and as disinfectants in the trenches. The element therefore prevented the spread of life-threatening infectious diseases such as cholera, typhoid, and typhus.

White phosphorus is another example. The solid ignites spontaneously in air forming a white smoke. It was used not only to fill incendiary shells and bombs but also as an obscurant chemical in smoke shells. Smoke shells were fired at enemy positions to create a fog-like blanket that smothered and confused the troops. On the other hand, obscurant smokes were also used to generate smoke screens for protective purposes. The purpose of the screens was to conceal military manoeuvres.

Steel, essentially an alloy of iron and carbon and sometimes other chemical elements, was used to make weapons and also for defensive purposes. Battleships and tanks relied on steel armour for protection. When the British began to issue front-line troops with steel helmets in 1915, battlefield head injuries dropped substantially. One British design, known as the Brodie helmet, was fabricated from a tough hard-wearing alloy steel containing manganese.

Photography, so important in the war, was another area that totally relied on chemistry. The history of photography can be traced back to the early 1800s and even before with the discovery of light-sensitive chemicals. By the beginning of the World War I, photography had come of age. Photographs and moving films provided enduring images of the war and brought home dramatic pictures of the horrors of the war often in vivid detail (Figure 1.5). Soldiers loved to pose in their uniforms for the camera and send postcards home. Photographs and motion pictures were used for official records, for propaganda purposes, for aerial reconnaissance, for training, for recruitment, and for coverage of the war in newspapers, magazines, and cinemas back home.

L.285. German and British wounded waiting to be evacuated.

Figure 1.5   The horrors of war: wounded British and German soldiers await transport at a casualty clearing station in northern France, March 1918 (Wellcome Library, London).

Every unexposed photographic plate or film was coated with an emulsion containing a photosensitive silver halide salt. The films were developed with chemical solutions and still photographs printed with inks composed of chemicals.

THE CARING ROLE

Care of the troops, sailors, airmen, and prisoners in the various theatres of engagement added another chemical dimension to the Great War. Troops in the front line, for example, often lived for days in trenches infested with fleas, flies, lice, rats, and unseen pathogenic microorganisms in the mud, soil, dirty water, and dead bodies. When possible and if available, the troops were supplied with disinfectants with bleaching properties, such as calcium hypochlorite or sodium hypochlorite, to sterilize drinking water and disinfect trenches and latrines.

Carbolic acid was widely used as a disinfectant and an antiseptic to treat wounds. The acid was one of many organic chemicals extracted from coal tar that were used in the war. Other antiseptic preparations included tincture of iodine and aqueous solutions of hydrogen peroxide.

A wide range of medicinal chemicals were employed to treat sick and wounded troops, although they were not always available in the casualty clearing stations and hospitals behind the front lines. Chloroform, ether, ethyl chloride, and nitrous oxide were all used as anaesthetics during surgery. Morphine, or morphia as it was also known, was widely used as a painkiller in the war.

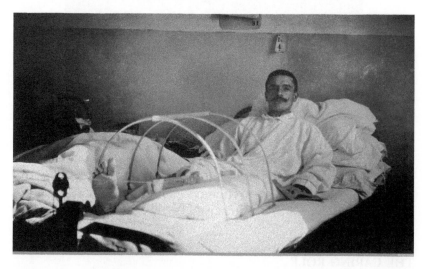

Figure 1.6 A military hospital in France. The soldier has had a leg amputated after fighting at Verdun in 1916 (Wellcome Library, London).

CHEMISTRY UNDERPINNED THE WAR EFFORT

As early as 1917, the First World War was becoming known as "The Chemists' War." Richard B. Pilcher (1874–1955), registrar and secretary of the Institute of Chemistry of Great Britain and Ireland, used the term in an article published in April that year.[3] He wrote: "In addition to competent chemical advisers of un-doubted standing, the following appear to be essential: Chemists to control the manufacture of munitions, explosives, metals, leather, rubber, oils, gases, food, drugs; chemists for the analysis of all such materials and for research; chemists on active service, to assist in the control of water supplies, in the detection of poison in streams, in the analysis of water and food, the disposal of sewage, and in other hygienic matters; chemists, both at home and in active service, to assist in devising safeguards against enemy contrivances of a scientific nature, and methods of of-fence to meet the same, as well as for the instruction of troops in such matters."

His list does not include, but might well have included the production of antiseptics, disinfectants, anaesthetics, natural and synthetic dyes and pigments, and photographic chemicals. Even so, it is clear that chemists, chemistry, and chemicals underpinned the war effort in the trenches, in the tunnelling operations, in the air and at sea, and in the casualty clearing stations and military hospitals. Chemistry was unquestionably the *sine qua non* of the war.

REFERENCES

1. W. D. Cocroft, First World War explosives manufacture: the British experience, in *Frontline and Factory: Comparative Perspectives on the Chemical Industry at War, 1914–1924*, ed. R. MacLeod and J. A. Johnson, Springer, Dordrecht, 2006, p. 36.
2. A. M. Prentiss, *Chemicals in War: A Treatise on Chemical Warfare*, McGraw-Hill, New York, 1937, p. 661.
3. R. B. Pilcher, Chemistry in wartime, *Ind. Eng. Chem.*, 1917, **9**, 411.

CHAPTER 2

Calling All Chemists

ALLIED CHEMISTS MOBILISE

The British Army "owes a great debt to science, and to the distinguished scientific men who placed their learning, and skill at the disposal of their country." The compliment came from Field Marshal Sir Douglas Haig (1861–1928) in his final despatch as Commander-in-Chief of the British Armies in France following the end of the First World War.[1] The despatch was dated 21 March 1919 and published the following month when Haig was appointed Commander-in-Chief of British Home Forces.

Without science, he suggested, the Allies could not have attained "general superiority" in the development of the "mechanical contrivances" which contributed so powerfully to Germany's defeat in the war. Haig referred specifically to "the development of motor transport, heavy artillery, trench mortars, machine guns, aeroplanes, tanks, gas and barbed wire." He also alluded to smoke shells, new fuses, aerial photography, anti-gas precautions, submarine warfare, and hospitals.

After the war of movement on the Western Front in August and September 1914, the conflict evolved into a stalemate "with creation of continuous trench lines from the Swiss frontier to the sea," Haig noted. The war became a war of not just men but also of materials, the materials being the guns and ammunition to

The Chemists' War, 1914–1918
By Michael Freemantle
© Freemantle, 2015
Published by the Royal Society of Chemistry, www.rsc.org

fight the enemy and the raw and intermediate materials required to manufacture the munitions.

The conflict therefore gradually developed into a war of attrition or a "wearing out struggle" as Haig called it. The British Royal Navy blockade of German imports played an important role that contributed to Germany running out of essential material resources towards the end of the war. "The rapid collapse of Germany's military powers in the latter half of 1918 was the logical outcome of the fighting of the previous two years. It would not have taken place but for that period of ceaseless attrition which used up the reserves of the German Armies while the constant and growing pressure of the blockade sapped with more deadly insistence from year to year at the strength and resolution of German people."

The First World War pitted the industrial might of the Central Powers, notably Germany, against the industrial might of Britain, France and the other Allied Powers resulting in slaughter and destruction on an industrial scale. But it wasn't the first industrial war. The American Civil War from 1861 to 1865 saw the more industrialised northern states of the Union gain victory over the southern states of the Confederacy with their agricultural slaved-based economy. The north, for example, had twice as many miles of railway track as the south and with its abundance of raw materials and technological expertise was able to ensure that the Union armies were well equipped.[2]

The Prussians and their German allies also exploited the latest advances in technology to win their war with France that was fought from 1870 to 1871 (the Franco–Prussian war, also known as the Franco–German war). The Prussian steel breech-loading field gun, for example, had far better accuracy, range and rate of fire than the French field gun, and the Prussians used their railways much more effectively and efficiently to mobilise their army and transport soldiers, munitions and other supplies to the front line.

The Great War, although not the first industrial war, was the first industrial *world* war and, like the previous industrial wars, it relied extensively on the mass production of weapons, ammunition and other war matériel, all of which, as Haig implied, was underpinned by engineering, technology and science, not least chemistry.

Britain made a shaky start in this respect. Many well-qualified chemists volunteered for service in the armed forces at the start of the war and a significant number of them were killed in action (see Chapter 19). The country was "unprepared for war, or at any rate for a war of such magnitude," Haig observed. "As regards material, it was not until midsummer, 1916, that the artillery situation became even approximately adequate to the conduct of major operations."

The chemists' war in the country had started in a muddle and in crisis, Roy MacLeod, a historian at the University of Sydney, Australia, pointed out in an article on the mobilisation of civilian chemists published in 1993.[3] In August 1914, the month Britain declared war on Germany, the British Board of Trade appointed a Chemical Products Supply Committee chaired by Edinburgh-born Richard Haldane (1856–1928), the country's Lord Chancellor and former Secretary of State for War. The War Office also set up an Explosives Supply Committee chaired by English lawyer John Moulton (1844–1921).

With the advent of trench warfare following the Battle of the Marne, France, in September 1914, industry was suddenly called upon to produce quantities of propellants and explosives on an unprecedented scale. The German introduction of gas as a lethal weapon at Langemarck, Belgium, on 22 April 1915 (see Chapter 13) upped the ante even further. The Secretary of State for War, Lord Horatio Herbert Kitchener (1850–1916), asked Scottish physiologist John Scott Haldane (1860–1936), the brother of Richard Haldane, to go to Belgium to "investigate the effects of the gas and to offer suggestions as to how the danger could be met."[4] John Haldane identified the gas as chlorine and then rushed back to England to speed up the production of an emergency respirator. He subsequently advocated the development of the box respirator that became the standard anti-gas equipment for troops. It was designed by British chemist Edward Harrison (1869–1918) and introduced by the British Army in February 1916 (see Chapter 19).

On 12 May 1915, Richard Pilcher at the Institute of Chemistry of Great Britain and Ireland received a notice from the War Office calling for "men with a knowledge of chemistry" which, "he believed," had also been sent to colleges and universities throughout the country.[5] "The purpose for which the chemists were required remained a secret ... and it was not until after the first gas attack that any definite information with regard to the force was available," he noted at the institute's 41st annual general meeting that took place in London on 3 March 1919.

In early May 1915, in retaliation to the German gas attack the previous month, Kitchener authorised the development and use of chemical weapons. In July and August that year, the British Army established four "Special Companies" to take responsibility for the use of poison gas. They were commanded by Lieutenant Colonel Charles Howard Foulkes (1875–1969) and included science graduates and industrial chemists. The companies carried out the first British gas attack at the Battle of Loos in September 1915 when they attacked the Germans with clouds of chlorine gas delivered from cylinders. At Haig's request, the War Office subsequently expanded the companies into a force known as the Special Brigade that consisted of over 5500 soldiers.

NAVVIES, TUNNELLERS AND CHEMISTS

Some chemists in Britain were unhappy at the way chemists were employed in the Special Brigade. The chemists "who were enlisted into it or were transferred from other units to it, were given work that could be done as well, or perhaps rather better, by plumbers' assistants," complained Lewis Eynon at the Institute of Chemistry's March 1919 meeting in London.[5] He regretted that the institute had not protested "against the scandalous waste of qualified chemists on unskilled labour ... by the military authorities." He added "that with the usual stupidity they took men who could have done much better in other work."

Cecil Jones, who also attended the meeting, agreed. Although it was too late to criticise the War Office, some "chemists were turned over to a job which no chemist would waste a laboratory boy on—a crossing sweeper could have done the work." Another speaker at the meeting, Dr Brady, said that he knew a considerable number of chemists in the Special Brigade and many felt "very sorely" on the point. He referred to a leaflet issued by the Parliamentary Recruiting Committee in which "volunteers for special corps were called for, and Grade 1 was, 'navvies, tunnellers and chemists'."

A report published by the institute in 1919 indicated that almost "800 members and students of the Institute, or rather more than 25 per cent" of the total membership were on active service during the war.[6] The numbers did not include members of the three British chemical societies (see Chapter 19) who were not members of the institute.

The Ministry of Munitions also asked Pilcher to find trained chemists to work in the country's munitions factories. Some chemists in the Special Brigade were brought back to Britain to carry out research and others were recruited from civilian employment. In March 1916, military conscription came into force under the Military Service Act. Before that, the British Army had been a volunteer army. Following pressure from the Royal Society, Britain's premier learned scientific society, a number of "war service badges" were issued to chemists, biologists, physicists, and metallurgists. The holders of these badges were exempt from national service because their work was considered to be of national importance for the war effort. By January 1918, 312 analytical and consulting chemists were included on the reserved occupations list and a total of 482 scientists and technicians worked for the ministry.[7]

Some 1460 scientists and assistants in Britain participated in chemical warfare research and development, noted Ludwig Haber (1921-2004),[8] the younger son of German chemist and Nobel laureate Fritz Haber (see Chapter 12). Many of these worked at "The War Department Experimental Ground," an experimental station set up by the Ministry of Munitions at Porton Down, Wiltshire, in March 1916 to study the scientific aspects of chemical warfare. The establishment was headed by Arthur Crossley (1869–1927), a professor of organic chemistry at King's College, London (see Chapter 18).

The total number of civilian chemists mobilised by Britain for its "chemical army" is not precisely known, according to MacLeod.[3] Whatever its size, the arsenal of "garrison chemists" in this army transformed the country's chemical industry and moved the munitions industries "from crisis to confidence," he noted.

By late 1916, Britain as well as Canada, France and Italy had transferred all chemists remaining in active military service to chemical duty at home. By then, "many had already been lost

Figure 2.1 Arthur Crossley (Royal Society of Chemistry Library).

and their loss was seriously felt," reported Charles Lathrop Parsons (1867–1954), chairman of the American Chemical Society (ACS) Committee on War Service for Chemists, at the society's 56th meeting in Cleveland, Ohio, on 10 September 1918.[9]

On 5 February 1917, two months before the United States declared war on Germany, ACS president Julius Stieglitz (1867–1937) sent a telegram to Woodrow Wilson (1856-1924), President of the United States, informing him that the society "with over eight thousand members, begs to place its services" at his command, "especially in matters facilitating preparations of munitions, supplies, medicinal remedies, and other chemical materials."[10] The president replied with "grateful thanks" for the "very generous and patriotic proffer" of the society's services.

In Russia, universities and technical schools carried out experimental work on the preparation of new drugs, vaccines, iodine and dyestuffs for the war effort while leading scientists at Moscow University investigated the properties of explosives and gases.[11] Documentary evidence about the mobilisation of chemists and other scientists in Russia and some other countries such as Italy is sparse, however.[12]

In France, specialist workers were hastily recalled from the army as early as September 1914 following the rapid decline in stocks of ammunition in the country. They helped to boost the manufacture of shells, explosives and other war matériel and, to show that they were not shirkers, each worker wore a red armband displaying a grenade.[13] A total of about 110 scientists and assistants worked on chemical warfare in France according to Ludwig Haber.[8] In his book on scientific developments in the First World War, British author Guy Hartcup reported that the French ministry responsible for armaments and munitions was employing almost 750 scientists and technicians in January 1918.[14]

It is important to add a caveat here. Ludwig Haber, who delved deeply into chemists-at-war statistics for the major belligerent countries remarked that there are "plenty of numbers, but few stand up to close scrutiny."[15] For instance, he put the number of scientists and assistants engaged in chemical warfare research and development in the United States at 1900. Parsons gave slightly different numbers at the Cleveland meeting. He noted that when the Chemical Warfare Service was set up in the United States in June 1918, "there were over 700 chemists at work on problems having to do with gas warfare, the design of gas masks, protection against toxic gases, development of new gases, and the working out of processes for those already used, the details of incendiary bombs, smoke funnels, smoke screens, smoke grenades, colored rockets, gas projectors and flame throwers, thermal methods for combating gas poison, gases for balloons, and other materials directly or indirectly connect with gas warfare." In addition there were "nearly 1100 helpers in the way of clerical force, electricians, glass blowers, engineers, mechanics, photographers, and laborers," making a total of 1800 employees.

GERMANY AHEAD IN 1914

The origin of modern chemistry as we know it is widely attributed to French chemist Antoine Laurent Lavoisier (1743–1794) and the publication of his chemistry textbook

Traité Élémentaire de Chimie (Elementary Treatise on Chemistry) in 1789. During the 19th century the development and application of chemistry progressed at a remarkable pace. The serendipitous discovery of mauve, the first synthetic organic dye, by English chemist William Henry Perkin (1838–1907) in 1856 proved to be a major landmark (see Chapter 5). He prepared the dye from aniline, an organic chemical derived from coal tar.

The Germans soon began to exploit the commercial potential of synthetic organic chemistry and by the time the First World War started, the country dominated the world market in the production and export of dyes, pharmaceuticals, and other synthetic organic chemical products. Both Britain and the United States, for example, imported around 90% of their dyes from Germany. These dyes were needed for their textile, wallpaper, printing, paint and other industries. Germany also exported vast quantities of drugs, disinfectants, analgesics, anaesthetics, and organic chemicals required for photography and the manufacture of scents and perfumes.

It is an irony of history that chemistry, procreated and born in France, nursed and educated in Great Britain, was at last widened and perfected by the latecomer Germany in such a way that by 1914 it was called the pharmacy of the world, remarks German chemist Walter Wetzel in an article about the German chemistry profession in the second half of the 19th century.[16]

For example, August Wilhelm von Hofmann (1818–1892), who was born in Giessen, Germany, has been widely credited with sparking the expansion of the chemical industry in the country. The German chemist, however, spent many years in Britain. He was the first director of Britain's Royal College of Chemistry that had been founded in 1845 and was the forerunner of the Department of Chemistry at Imperial College London. Perkin entered the college in 1853 and studied chemistry under Hofmann. Hofmann appointed him as an assistant in 1855 and assigned him the project that led to the discovery of mauve. Hofmann subsequently discovered a way of synthesising fuchsine, a magenta dye, from aniline.

Hofmann returned to Germany to become professor of chemistry at Berlin University in 1865. He took with him his knowledge and expertise in coal tar chemistry and within a few years the coal-tar dyestuff industry in Germany began to

Figure 2.2   August Wilhelm von Hofmann (Wellcome Library, London).

blossom. By 1906, some 4500 chemists worked in the German chemical industry compared with just 500 chemists in the British chemical industry, notes historian Michael Sanderson.[17] When the First World War started, German chemical industry employed one university-trained chemist for every 15 workers in the industry.[18] German industry as a whole employed one such chemist for every forty workers whereas in Britain, the ratio was one university-trained chemist for every 500 workers.

The Kaiser Wilhelm Society for the Advancement of Science founded in 1911 provided strong links between academic science and industry. Two of the society's institutes were established in Dahlem, Berlin, that year: one for chemistry and the other for physical chemistry and electrochemistry (see Chapter 12).

The Kaiser Wilhelm Institute for Coal Research, founded in Mülheim the following year, further strengthened the industrial application of chemistry in the country.

In the final year of the First World War, Germany was able to employ some 2000 scientists and assistants in chemical warfare research and development—more than Britain and France combined.[8] The Austro-Hungarian Empire on the other hand was far less enthusiastic about the development and application of scientific research for its war effort. The empire "lacked the drive that produced the great advances in German science and technology," according to Hartcup.[19]

ALARM BELLS IN BRITAIN

Early on in the war, British scientists were beginning to sound alarm bells about Germany's industrial prowess and Britain's inability to keep pace with developments in the German chemical industry. William Henry Perkin Jr. (1860–1929), whose father discovered mauve, was one of them. When the war started, he was professor of chemistry at Oxford University and president of Britain's Chemical Society. The "decadence" of the coal tar industry in Britain and "its gradual transference to Germany may be said to have begun during the period 1870–75," he observed in his presidential address on "the position of the organic chemical industry" delivered at the society's annual general meeting on 25 March 1915.[20]

"It is surely remarkable that the study of so important a subject as organic chemistry should not only have been practically ignored by our universities in the past, but that even at the present day it does not flourish in the way it does in almost every university and technical school in Germany," Perkin complained. The majority of German professors stay in close touch with industry and spend part of their time solving technical problems, he explained.

The manufacturer in Britain is "usually merely a commercial person who does not like the expert" and in many cases is ignorant of science and its value, he added. In Germany, on the other hand, "chemical experts" are paid salaries "quite unheard of in this country" and "business men understand that the control of a chemical works must be in the hands of the chemist."

Figure 2.3   William Henry Perkin Jr. (Royal Society of Chemistry Library).

In the spring of 1915, the British government established British Dyes Ltd in an attempt to jump-start the aniline-dye industry from "a history of vested interests, inbred trade practices, and invalid beliefs," MacLeod noted. "Remarkably, chemists were kept out of management."

Later that year, Oxford University professor of zoology Edward Poulton (1856–1943) echoed Perkin's comments concerning the failure of the coal tar industry in Britain. In a public lecture on "Science and the Great War" delivered in Oxford on 7 December 1915, he remarked that "the failures which have occurred are nearly all due to the national neglect of science and the excessive predominance in Parliament, and especially in the Government, of the spirit that is most antagonistic to science—the spirit of the advocate."[21]

Poulton referred to the "grave responsibility" incurred by the government in deciding vital issues without scientific evidence. "Untold thousands of lives and an ever-growing volume of human misery are a terrible punishment for the neglect of science," he said. "It is possible that military experts are mistaken in thinking that the final decision can be reached by fighting. It may have to be reached by economic and financial pressure."

REFERENCES

1. D. Haig, Final Despatch. Part II: Features of the War, *The London Gazette* (4th Supplement), 8 April 1919.
2. A. Farmer, *The American Civil War: Causes, Course and Consequences 1803–77*, 4th Edn, Hodder Education, London, 2006, p. 13 and p. 133.
3. R. MacLeod, The chemists go to war: mobilisation of civilian chemists and the British war effort, 1914–1918, *Ann. Sci*, 1993, **50**, 455.
4. C. G. Douglas, John Scott Haldane. 1860–1936, *Obit. Not. Fell. R. Soc.*, 1936, **2**, 114.
5. Anon., Proceedings of the 41st Annual General Meeting of the Institute of Chemistry of Great Britain and Ireland. Friday, March 3rd, 1919, *Proc. Inst. Chem. GB Irel.*, 1919, **43**, B001.
6. Anon., Ex-service chemists, *Proc. Inst. Chem. GB Irel.*, 1919, **43**, D29.
7. G. Hartcup, *The War of Invention: Scientific Developments, 1914–18*, Brassey's Defence Publishers, London, 1988, p. 23 and p. 31.
8. L. F. Haber, *The Poisonous Cloud, Chemical Warfare in the First World War*, Clarendon Press, Oxford, 1986, p. 107.
9. C. L. Parsons, The American chemist in warfare, *Ind. Eng. Chem.*, 1918, **10**, 776.
10. Anon., Editorial Chemists and the country's crisis, *Ind. Eng. Chem.*, 1917, **9**, 224.
11. P. Gatrell, *Russia's First World War: A Social and Economic History*, Pearson Education, Harlow, 2005, p. 48.
12. G. Hartcup, *The War of Invention: Scientific Developments, 1914–18*, Brassey's Defence Publishers, London, 1988, p. 33.
13. M. Hastings, *Catastrophe: Europe Goes to War 1914*, William Collins, London, 2013, p. 417.
14. G. Hartcup, *The War of Invention: Scientific Developments, 1914–18*, Brassey's Defence Publishers, London, 1988, p. 31.

15. L. F. Haber, *The Poisonous Cloud, Chemical Warfare in the First World War*, Clarendon Press, Oxford, 1986, p. 341.
16. W. Wetzel, Origins of and education and career opportunities for the profession of 'chemist' in the second half of the nineteenth century in Germany, in *The Making of the Chemist: The Social History of Chemistry in Europe 1789–1914*, ed. D. Knight and H. Kragh, Cambridge University Press, Cambridge, 1998, p. 77.
17. M. Sanderson, The University of London and industrial progress 1880–1914, *J. Contemp. Hist.*, 1972, 7, 243.
18. W. R. Whitney, England's tardy recognition of applied science, *Ind. Eng. Chem.*, 1915, 7, 819.
19. G. Hartcup, *The War of Invention: Scientific Developments, 1914–18*, Brassey's Defence Publishers, London, 1988, p. 36.
20. W. H. Perkin, Presidential address, the position of the organic chemical industry, *J. Chem. Soc., Trans.*, 1915, **107**, 557.
21. E. B. Poulton, Science and the Great War, The Romanes Lecture, 1915, Oxford University Press, Oxford, 1915.

CHAPTER 3

Women's Contributions

TOTAL WAR

"In the end ... it is impossible for us to lay the responsibility of warlike operations merely upon the soldier," concluded English chemist Sir William Jackson Pope (1870–1939) in an address on "modern developments in war making" that he delivered at the 62nd meeting of the American Chemical Society (ACS) held at Columbia University in September 1921.[1] "We are all of us responsible. The chemist has been largely responsible, the medical man has been responsible; the man who bought the German dyes has been responsible, and the man who contributes to war funds; we are all of us in the same boat."

That same boat was the boat of "total war," that is a war where a combatant nation employs all its available resources to fight the war. The resources include members of the armed forces and civilians whether male or female, their experience, their know-how, their skills and expertise, and their physical abilities. They include land, natural resources such as wood and minerals, farms, roads, railways, airfields and aircraft, ships and ship-yards, factories, and other infrastructure. As far as chemistry was concerned, the total war of 1914–1918 more or less involved every chemist in one way or another; all available chemicals;

The Chemists' War, 1914–1918
By Michael Freemantle
© Freemantle, 2015
Published by the Royal Society of Chemistry, www.rsc.org

every academic, government and industrial chemistry laboratory; every last test-tube and Bunsen burner; and the entire chemical industry.

Wherever possible, everyone and everything in the nation's boat was directed to the war effort. In Britain during the First World War, boy scouts served as lookouts with coastguards to help protect the English coast and its ports and estuaries from enemy invasion. Men who were infirm, unfit or too old to enlist in the armed forces enrolled as special constables to man road blocks, and women replaced the farm, factory, railway and other male workers who volunteered for or were conscripted into the armed services.

As the war progressed, British women took on jobs that had previously been considered unsuitable for women. Women became chimney sweeps, road sweepers, bicycle messengers, park keepers, gardeners, funeral directors, bill-posters, window cleaners, milk deliverers, bricklayers, coal heavers, and also worked in breweries and foundries and as stokers in gasworks.[2] Other women became tram conductors, gamekeepers, electric train cleaners, potato shovellers, leather dippers, steam roller drivers, pork butchers, and refuse collectors.[3]

The number of women in employment mushroomed. Whereas there were just one thousand woman railway clerks in 1914, there were 14 000 in 1918.[4] Overall, the number of women employed in transport grew from 18 000 in 1914 to 117 000 in 1918.[5] Tens of thousands of British women volunteered to join organisations such as the Women's Land Army which provided labour for the farming industry or the Women's Forestry Corps which delivered timber to build trenches, make railway sleepers, fabricate ammunition packing cases, and to meet other needs of the armed forces.

Substitution of women for men in the workforce occurred throughout the belligerent countries in Europe. In French cities, for example, women delivered the post and were also employed as firefighters.[6] In Russia, the proportion of women in the manufacturing industry labour force rose from 30% in 1913 to 40% in 1916.[7] The number of women in Russian engineering trades increased from 23 000 to 98 000 over the same period.

Figure 3.1   Motor ambulances with their women drivers at Étaples, France, 1918 (Wellcome Library, London).

MUNITIONS FACTORIES

As with so many other statistics of the First World War, sources differ on the number of women who worked in munitions factories during the war. According to British historians Neil Storey and Molly Housego, some 75 000 British women were employed in the munitions industry producing millions of shells and small arms ammunition each year.[8] Arthur Marwick, another British historian, noted that just before the outbreak of war, "there had been 212 000 women employed in what were to become the munitions industries."[5] By July 1916, the number had risen to 520 000 and by the end war it had risen to over 900 000. He added that the total workforce of "women and girls over ten" in industry as a whole increased by around 800 000 from almost 2.2 million in 1914 to almost 3.0 million in 1918.

The numbers of "married women of child-bearing age" working in British industry steadily increased throughout the war, noted a report on expectant mothers in a munitions factory

Figure 3.2   The oil painting "The Munitions Girls" by Alexander Stanhope
Forbes shows women working at Kilnhurst Steelworks,
Rotherham, England, during the First World War (Science
Museum, London, Wellcome Library Images).

published in September 1918.[9] "The question of the employment of the working woman who becomes pregnant has always
been a vexed one," the author commented. Their report described a "prematernity scheme" at national ordnance factories
in Leeds whereby expectant mothers who worked nights and
carried out heavy lifting were transferred to day shifts and light
sedentary work at the end of the fourth month of pregnancy.
They assembled fuses for shells from 7.30 a.m. to 5.30 p.m. "with
the usual factory breaks for meals." After seven months of
pregnancy, they were engaged in making and mending overalls,
caps, trouser suits, gloves and other items of clothing for the
factory workers. "The last hour of the day is given up to the
making of clothes for the baby under the help and guidance of
the sewing dépôt forewoman."

Female workers in British munitions factories became known as "munitionettes." Some were killed or injured in explosions such as the Silvertown explosion on 19 January 1917 when 50 tons of TNT exploded (see Chapter 19). Munitionettes also experienced health problems arising from the exposure to toxic chemicals. For example, when they came into contact with TNT, a yellow powder, their skin became yellow. For this reason, they were known as "canaries" or "canary girls." Prolonged exposure to the explosive caused headaches, loss of appetite, vomiting, many other symptoms, and eventually death.[10]

CARE OF THE SICK AND WOUNDED

Many women in Great Britain and Ireland volunteered for nursing in the war. The British Red Cross and St John Ambulance established the Voluntary Aid Detachments (VADs), which provided nursing services in the field under the protection of the Red Cross. They recruited and trained thousands of women as VADs. Thousands of trained uniformed nurses also served in the Queen Alexandra's Imperial Military Nursing Service, the nursing wing of the British Army, or the Queen Alexandra's Royal Naval Nursing Service. By 1918, more than ten thousand nurses served in these and other British military nursing services.[11]

Nurses were at risk of injury or death whether serving in casualty clearing stations near the front lines, field hospitals, military hospitals, or on hospital ships. When a military hospital in Étaples, northern France, was bombed on the night of 30–31 May 1918, for example, one ward was "blown to pieces" and others ruined or severely damaged.[12] "Sister Baines, four orderlies, and eleven patients were killed outright, whilst two doctors, five sisters, and many orderlies and patients were wounded."

Ten of the 36 New Zealand Army Nursing Service staff nurses on board HMT (His Majesty's Troopship) *Marquette* drowned when the ship was torpedoed and sank in the Aegean Sea on 23 October 1915. The ship was bound for Salonika, Greece, with a complement of 741 crew, nurses and troops. Altogether 167 lives were lost. As the vessel sank, there were many sad scenes, the worst being an accident to a lifeboat resulting in many nurses

Figure 3.3   Treatment of a wounded soldier at an advanced dressing station (Wellcome Library, London).

"being emptied into the sea" according to an account by one of the survivors who was "sucked nearly under the propeller."[13]

The war required the mobilisation of not just nurses to care for the sick and wounded but also doctors. A substantial proportion of the medical profession in Britain were called up for military service in the Royal Army Medical Corps (RAMC) which had been founded in 1898. Over the course of the war the number of officers and men in the corps rose from just under 5000 to well over 144 000.[14] With so many medical men leaving the country for active service overseas, there was growing recognition that women doctors would be needed to replace them at home.[15] The London teaching hospitals started to admit women and the number of civilian opportunities for qualified women doctors began to increase.

Even so, a number of British women doctors volunteered for war work abroad. One of them was Florence Stoney (1870–1932), a pioneer of radiography who had 13 years of experience of X-ray treatment in English hospitals before the war.[16] In September 1914, she was appointed chief medical officer of an all-women medical unit that worked with the Belgian Red Cross to set up a field hospital in an abandoned music hall in Antwerp, Belgium. The unit had been formed by Mabel St Clair Stobart (1862–1954), a non-medical woman who had founded the Women's Sick and Wounded Convoy Corps and the Women's National Service League in 1912 and 1914, respectively.

When German artillery bombarded Antwerp in October 1914 the music hall was one of the first buildings to be shelled. Stoney, her staff, and all the wounded managed to avoid capture and escaped to Holland which remained neutral in the war. They left behind their equipment including X-ray apparatus, an operating table, surgical instruments, and sterilisers. Stobart quickly raised funds to purchase new equipment and set up a hospital for the French Red Cross in a 16th century castle near Cherbourg. Stoney's team successfully treated over 100 severely injured soldiers even though the hospital's electricity and water supplies were primitive. The hospital closed in March 1915 and, at the request of the British War Office, Stoney returned to England to take charge of the X-ray department at Fulham Military Hospital near London. Stobart subsequently went to the Balkan Front where she led the Serbian Relief Fund's Front Line Field Hospital. She and her team of medical staff accompanied the Serbian Army and tens of thousands of civilians as they retreated from the attacking Central Powers' armies in bad weather and along poor roads through Montenegro to Albania. The retreat continued from October 1915 to February 1916.

WOMEN CHEMISTS

The total war of 1914–1918 required the mobilisation of women chemists, many of whom were employed in schools and universities. In the second half of the 19th century, women chemists in Britain, as elsewhere in the world, were few and far between compared with their male counterparts. But then, in the years leading up to the start of the First World War, the number of British women taking degrees in chemistry increased significantly, observed chemists Geoff Rayner-Canham and Marelene Rayner-Canham at Sir Wilfred Grenfell College, Memorial

University of Newfoundland, Canada. The Rayner-Canhams have carried out extensive studies of British women chemists in the war.[17] If it had not been for this pool of qualified women chemists who replaced the male chemists who had signed on for active service, "the British war machine would have faced a severe shortage of chemists," they point out.

Statistics for the University of Aberdeen, Scotland, reported by the Rayner-Canhams underline the point.[18] In the three years from 1911 to 1914, there were 74 male "chemistry enrolments" at the university compared with 29 female enrolments. For the four years of the war, 43 males and 58 females enrolled and then in the following four years from 1918 to 1922 the male and female numbers were 83 and 48 respectively.

The Rayner-Canhams comment that before the war, "women were tolerated in male professions" and perceived as curiosities rather than threats. During the war, professional women carried out a "significant portion" of skilled work and then after the war "discrimination against women" increased. Male professionals regarded them with "open hostility" because they presented a challenge to their own jobs.

Women chemists, their contributions to science, and their role in the war were also ignored. As we saw in the first sentence of the previous chapter, Sir Douglas Haig stated that the British Army owed a "great debt" to "distinguished scientific men" for Germany's defeat in the war. And Pope did not refer to women when he spoke about responsibility in war making. The quotes at the beginning of this chapter, in which he mentions "man" three times, are from the concluding paragraph of the published version of his address to the ACS in 1921. The following is the opening paragraph of that address: "Every scientific man is becoming more and more convinced that war, if it cannot be abolished, must at least be reduced to a minimum. Every scientific man has become more and more convinced, during the last seven years, that the future of war depends upon science; and practically every scientific man, at least in Great Britain and France, has had seven years cut out of his life and out of his activities in research work by the events of the last seven or eight years."

The term "scientific man" appears three times and the word "his" twice. It was, he claimed, "scientific man" who sacrificed "his" research work for the war effort. Yet other male chemists of the time did acknowledge the existence and role of women chemists, albeit somewhat patronisingly at times.

"Women are coming into our profession—not in great numbers, it is true—but certainly sufficient for us to recognise that they are with us; and we know they are capable of doing excellent work," remarked Sir Herbert Jackson (1863–1936), professor of chemistry at King's College, London, in his presidential address at the 41st annual general meeting of the Institute of Chemistry of Great Britain and Ireland held in London on 3 March 1919.[19] "The Institute has admitted women chemists to its membership for nearly thirty years, and we look to our women chemists also to take their part in promoting the interests of the profession as a whole. We should not be properly representative without them, and it may be predicted that in some departments of chemical work we are likely to see them employed in greater numbers than they have been in the past."

The woman chemist has come to stay, announced an editorial in an issue of ACS's *Journal of Industrial and Engineering Chemistry* published the same month.[20] "War brought woman into industrial chemistry. The draft of the men of the laboratories into war service made necessary her increased employment. Many were the misgivings, but these have proved groundless."

"Perhaps the most interesting testimony is offered by Mr. William M. Brady, chief chemist of the Illinois Steel Company, in a statement published in the February number of *The Chicago Chemical Bulletin*," the editorial continued. Concerning the women in his laboratory, Brady is quoted as saying: "They learned the work as quickly as any men of like training could have learned it." He points out that women took less sick leave than men and adds patronisingly: "They have added tone to our laboratory by their pleasing personalities. They have proved beyond a doubt that they can do and will do at any hour of the day or night, careful, conscientious, reliable chemical work. They have passed the crucial test of service. They have been weighed in the chemical balance and not found wanting."

WOUNDED BY MUSTARD GAS

In Britain, women chemists contributed to the war effort in a variety of ways, not least in the production of fine chemicals in university and industrial laboratories.[21] English chemist Martha Annie Whiteley (1856–1956) was a notable example. She received a doctor of science degree from the University of London in 1902. Two years later, she was appointed chemistry assistant at Imperial College London and in the same year was one of 19 women chemists who signed a petition appealing to the president and council of the Chemical Society to allow qualified women chemists to be admitted into the fellowship of the society. The appeal failed and it was not until 1920 that Whiteley and 20 other women chemists were admitted to the society.

Figure 3.4   Martha Whiteley (courtesy: Archives Imperial College, London).

In the years leading up to the war, Whiteley carried out research on organic acids and related compounds. She "provided an every-ready source of inspiration and guidance for her research students and junior colleagues in the laboratories for organic chemistry," noted her Chemical Society obituarist.[22] By the time war broke out, Whiteley had been promoted to chemistry lecturer at Imperial College.

The war brought shortages of synthetic and naturally-occurring pharmaceuticals, anaesthetics, dyes and other fine chemicals that Britain had previously imported in considerable quantities from Germany. The British government tackled the problem by embarking on a crash programme aimed at producing the fine chemicals needed for its war machine. It enlisted the assistance of university and technical college chemistry departments not only to find and develop processes for the manufacture of these chemicals but also to make them in their laboratories.

Whiteley gathered together a team of women chemists and collaborated with Jocelyn Field Thorpe (1872–1940), professor of organic chemistry at Imperial College, on the synthesis of anaesthetics and drugs needed for military hospitals.[23] They included the local anaesthetics beta-eucaine and novocaine, and the analgesic phenacetin. In addition, Whiteley carried out research for the chemical warfare department of the Ministry of Munitions on tear gases and blister agents.

The Imperial College chemists also analysed samples of flares, explosives, and chemical warfare agents collected from the battlefields or from bombed areas in Britain. They were "very exciting days," observed Whiteley in a speech made in 1953.[24] She even examined a sample of the mustard gas that was "used to such effect on the front that our troops had to evacuate from Armentières as it was reputed to cause blisters."

The German artillery bombarded the French town with mustard gas shells in April 1918. To test the vesicant properties of the oily liquid, Whiteley applied a tiny smear to her arm and, she said, "for nearly three months suffered great discomfort from the widespread open wound it caused in the bend of the elbow, and which I still carry the scar." She added that shortly afterwards, when her team was conducting research on a method of making mustard gas, "my arm was always in requisition for the final test."

Whiteley was elected a Fellow of the Royal Institute of Chemistry in 1918 and two years later appointed an Officer of the Order of the British Empire (OBE) in recognition of her services to the nation during the war. She was the co-author with Thorpe of *A Students' Manual of Organic Chemical Analysis: Qualitative and Quantitative* which was published in 1925.[25] In 1928, she was the first woman to be elected a member of the Council of the Chemical Society. Thorpe was knighted in 1939, the year before he died.

THE FIRST ROYAL SOCIETY FELLOWS

Many of the British women chemists in the early 20th century "veered towards biochemistry," according to the Rayner-Canhams.[26] One such chemist was Marjory Stephenson (1885–1948) who was to play an active role in the First World War, although not as a chemist. She was born near Cambridge and from 1903 to 1906 studied science at Newnham College, a women-only college at Cambridge University. The science consisted of chemistry, physiology and zoology. In those days, women students wore long skirts and had to be chaperoned when attending lectures outside the college. Practical work on biology was carried out in the university's Balfour Biological Laboratory for Women. Newnham had its own chemistry laboratory. "Only physiology was taught in the university laboratories at this date and women were excluded from the university classes in the other two subjects," noted Glasgow-born zoologist Muriel Robertson (1883–1973) in an obituary of Stephenson.[27]

From 1906 to 1911 Stephenson taught domestic science at colleges in Gloucester and London and then took up an appointment as an assistant in a biochemistry laboratory at University College, London. She gave classes on the biochemistry of nutrition and worked on a variety of biochemical research projects including studies of lactase, a digestive enzyme, and the metabolism of diabetes. The outbreak of war interrupted her tenure of a fellowship for medical research that she had been awarded the previous year. In October 1914, she joined the British Red Cross and travelled to France where she supervised hospital kitchens. She later served in Salonika where, according to Robertson, "she was in charge of a nurses convalescent home and also had some responsibility for invalid diets."

Figure 3.5   Women-only chemistry laboratory at Newnham College, Cambridge (courtesy: The Principal and Fellows, Newnham College, Cambridge).

Immediately after the war, she was honoured with an MBE (Member of the Most Excellent Order of the British Empire) for her services in the war. In early 1919, she moved back to Cambridge and resumed her fellowship in the university's biochemistry laboratory. She remained at Cambridge for the rest of her life carrying out research on chemical microbiology, most notably on bacterial metabolism. Her book on the topic was published in 1930.[28]

In 1945, Stephenson and British crystallographer Kathleen Lonsdale (1903–1971) were elected as Fellows of The Royal Society, Britain's most prestigious learned scientific society. They were the first two woman scientists to receive the honour since its foundation in1660. Stephenson and Lonsdale were elected into the Division B (biological sciences) and Division A (physical sciences) of the society, respectively.

Lonsdale was eleven years old when the First World War broke out. She was living with her family in Essex when she witnessed zeppelin bombing raids on London and saw one of the German airships shot down in a ball of flames.[29] She became a pacifist

Figure 3.6 Marjory Stephenson (courtesy: The Principal and Fellows, Newnham College, Cambridge).

and in the Second World War was jailed for one month in London's Holloway Prison for refusing to register for civil defence duties.

Robertson, Stephenson's Royal Society obituarist, worked at the Lister Institute of Preventive Medicine, London, on the problem of gas gangrene during both world wars.[30] Gas gangrene is the death and decay of wounded body tissue infected by several species of clostridial bacteria, in particular *Clostridium perfringens*, found in the soil. The infection results in the release of toxins and the production of gases inside the infected body tissue.

Figure 3.7   Muriel Robertson fishing for leeches (Wellcome Library, London).

Robertson's work on gas gangrene clostridia in the First World War focused on preparing an antitoxin for the treatment of the infection, "an aim realized only by the end of the war."[31] She also worked on typhus, a louse-borne disease that was endemic on the Eastern Front in the war, and on the development of a tetanus antitoxin. Tetanus, also known as "lockjaw" is an infectious disease caused by *Clostridium tetani*, a species of clostridial bacteria found in soil, manure and even dust contaminated with faecal matter. The death rate among tetanus cases in home military hospitals in Britain fell from almost 60% at the beginning of the war to about 20% from June 1917 onwards, according to a report published in 1918.[32] The report attributed the drop to the introduction of prophylactic tetanus antitoxin injections in late 1914.

The fact that Robertson was one of "the small band of pioneer women scientists in a not-too-appreciative masculine world" turned her into an "assertive and over-caustic critic of the good and the great of her time," wrote her Royal Society obituarists.[31] She was "highly critical" of the society although pleased when

Stephenson and Lonsdale were elected as fellows in 1945. "She was immensely pleased with her own election to the fellowship in 1947."

REFERENCES

1. W. J. Pope, Modern developments in war making, *Ind. Eng. Chem.*, 1921, **13**, 874.
2. N. R. Storey and M. Housego, *Women in the First World War*, Shire Publications, Oxford, 2010, p. 34.
3. K. Adie, *Fighting on the Home Front: The Legacy of Women in World War One*, Hodder & Stoughton, London, 2013, p. 45.
4. M. Hastings, *Catastrophe: Europe Goes to War 1914*, William Collins, London, 2013, p. 413.
5. A. Marwick, *The Deluge: British Society and the First World War*, Student Edition, Macmillan Press, London and Basingstoke, 1973, p. 91.
6. M. Hastings, *Catastrophe: Europe Goes to War 1914*, William Collins, London, 2013, p. 429.
7. P. Gatrell, *Russia's First World War: A Social and Economic History*, Pearson Education, Harlow, 2005, p. 67.
8. N. R. Storey and M. Housego, *Women in the First World War*, Shire Publications, Oxford, 2010, p. 38.
9. R. H. B. Adamson and H. Palmer-Jones, The work of a department for employing expectant mothers in a munition factory, *Br. Med. J.*, 1918, **2**, 309.
10. H. Thursfield, Note upon a case of jaundice from trinitrotuol poisoning, *Br. Med. J.*, 1916, **2**, 619.
11. N. R. Storey and M. Housego, *Women in the First World War*, Shire Publications, Oxford, 2010, p. 15.
12. N. R. Storey and M. Housego, *Women in the First World War*, Shire Publications, Oxford, 2010, p. 18.
13. A. Prentice, *Marquette disaster, Marlborough Express*, **L(2)**, 4 January 1916, p. 3.
14. N. A. Martin, Sir Alfred Keogh and Sir Harold Gillies: their contribution to reconstructive surgery, *J. R. Army Med. Corps*, 2006, **152**, 27.
15. L. Leneman, Medical women at war, 1914–1918, *Medical History*, 1994, **38**, 160.
16. Anon., Obituary: Florence A. Stoney, O.B.E., M.D., *Br. Med. J.*, 1932, **2**, 734.

17. M. F. Rayner-Canham and G. W. Rayner-Canham, British women chemists and the First World War, *Bull. Hist. Chem.*, 1999, **23**, 20.
18. M. F. Rayner-Canham and G. W. Rayner-Canham, Women in chemistry: participation during the early 20th century, *J. Chem. Educ.*, 1996, **73**, 203.
19. H. Jackson, The address of the president, *Proc. Inst. Chem. GB Irel.*, 1919, **43**, B0020.
20. Anon., The woman chemist has come to stay, *Ind. Eng. Chem.*, 1919, **11**, 183.
21. M. Rayner-Canham and G. Rayner-Canham, *Chemistry Was Their Life: Pioneer British Women Chemists, 1880–1949*, Imperial College Press, London, 2008, p. 8.
22. A. A. Eldridge, Martha Annie Whiteley: 1866–1956, *Proc. Chem. Soc.*, 1957, 157.
23. M. R. S. Creese, Martha Annie Whiteley (1866–1956): chemist and editor, *Bull. Hist. Chem.*, 1997, **20**, 42.
24. Anon., Speech by Whiteley at the Summer Luncheon of the Royal Holloway College Association Summer Luncheon, Royal Holloway College Association, College Letter 48–49, in M. Rayner-Canham and G. Rayner-Canham, *Chemistry Was Their Life: Pioneer British Women Chemists, 1880–1949*, Imperial College Press, London, 2008, p. 455.
25. J. F. Thorpe and M. A. Whiteley, *A Students' Manual of Organic Chemical Analysis: Qualitative and Quantitative*, Longmans, Green and Co., London, 1925.
26. M. Rayner-Canham and G. Rayner-Canham, Fight for rights, *Chemistry World*, March 2009, p. 56.
27. M. Robertson, Marjory Stephenson: 1885–1948, *Obit. Not. Fell. R. Soc.*, 1949, **6**, 562.
28. M. Stephenson, *Bacterial Metabolism*, Longmans, Green and Co., London, 1930.
29. M. Rayner-Canham and G. Rayner-Canham, *Chemistry Was Their Life: Pioneer British Women Chemists, 1880–1949*, Imperial College Press, London, 2008, p. 345.
30. J. Howie, Muriel Robertson, FRS (1883–1973), *Br. Med. J.*, 1987, **295**, 41.
31. A. Bishop and A. Miles, Muriel Robertson, 1883–1973, *Biogr. Mems Fell. R. Soc.*, 1974, **20**, 316.
32. Anon., Tetanus in home military hospitals, *Br. Med J.*, 1918, 2, 415.

CHAPTER 4

Nobel War Efforts

CENTRAL POWERS

It may come as a surprise to many readers that a significant number of chemists from both the Central Powers and the Allied Powers who contributed their services to the Great War in one way or another also won Nobel Prizes in Chemistry.

Some played relatively minor roles. Czech chemist Jaroslav Heyrovský (1890–1967) is one example. Born in Prague, then a city in Austria-Hungary, he began studying chemistry, physics, and mathematics in 1909 at the city's Czech University, later known as Charles University. The following year he transferred to University College, London, to continue his studies and graduated with a Bachelor of Science degree there in 1913. He enjoyed living in England during this period and was thinking of becoming a naturalised British subject. However, when war broke out, he was visiting his parents in Prague and was unable to return to England.[1]

In January 1915, he was conscripted into the Austro-Hungarian Army but because he was physically weak, the army posted him to a military hospital in Prague where he served as a dispensing chemist and radiologist throughout the war. His duties at the hospital were not onerous and so, at the same time, he managed to carry out research at the university in Prague on

The Chemists' War, 1914–1918
By Michael Freemantle
© Freemantle, 2015
Published by the Royal Society of Chemistry, www.rsc.org

his favourite branch of chemistry—electrochemistry. He was awarded a PhD degree there in 1918.

Four years later, he invented polarography, a powerful electrochemical technique that became widely used in chemical analysis.[2] In 1926, he was appointed professor of physical chemistry at Charles University where he continued developing the polarographic technique although his work was little known until the publication of his first book on the topic in 1933. He was awarded the Nobel Prize in Chemistry in 1959 "for his discovery and development of the polarographic methods of analysis."

Like Heyrovský, German chemist Eduard Buchner (1860–1917) also served in a military hospital during the war. Buchner won the 1907 Nobel Prize in Chemistry for his research on the fermentation of sugar. He should not be confused with Ernst Büchner (1850–1925), the German chemist who invented the Büchner vacuum filtration flask and funnel. Eduard Buchner was born in Munich and studied chemistry under Adolf von Baeyer (1835–1917), chemistry professor at the University of Munich. Baeyer was to win the Nobel Prize in Chemistry in 1905 for his research on organic chemistry.

Buchner obtained his doctorate at the university in 1888. He was appointed assistant lecturer in Baeyer's department the following year and two years later he was promoted to lecturer. From 1893 until the outbreak of the First World War, he conducted research programmes and taught chemistry at colleges and universities successively in Kiel, Tübingen, Berlin, Breslau and Würzburg. In the war, he served as a major in the German Army and was posted to a hospital at Focşani, a city in Romania, during the Romanian campaign in which the armies of the Central Powers fought against the Romanian and Russian armies from August 1916 to December 1917. He was seriously injured in a grenade attack on 3 August 1917 and died of his wounds nine days later at the age of 57.

Whereas Heyrovský's and Buchner's activities as chemists had little impact on the war, those of German chemist Emil Fischer (1852–1919) were hugely significant. Born in Euskirchen, Fischer was another future Nobel laureate who studied under Baeyer. He carried out numerous investigations into the chemistry of carbohydrates, amino acids, peptides, proteins, enzymes, purines (nitrogen-containing organic compounds with two-ring molecular structures) and many other organic compounds. His two-dimensional diagrams, known as Fischer projections, of the structures of glucose and other simple carbohydrates are found in numerous chemistry textbooks. He was the first German chemist to be awarded the Nobel Prize in Chemistry, winning it in 1902 for his work on carbohydrates and purines.

Fischer was appointed professor of chemistry at the University of Berlin in 1892 and developed close ties with the German chemical industry. He had political influence, notably with German emperor Kaiser Wilhelm II (1859–1941) and played a

Figure 4.1   Emil Fischer around 1900 at Friedrich Wilhelm University, Berlin (courtesy: Archiv der Max-Planck-Gesellschaft, Berlin-Dahlem).

leading role in the establishment in 1911 of the Kaiser Wilhelm Society for the Advancement of the Sciences and two Kaiser Wilhelm institutes.

By the time the war started, Fischer had become a towering figure in German chemistry and was therefore in an ideal position to make a substantial contribution to the German war effort. He abandoned most of his academic research and concentrated on serving as "a scientific intermediary between the military bureaucracy and industry."[3] His wartime activities were extensive throughout the conflict. For example, in December 1914, months before the first German chlorine attack, he had conversations with Erich von Falkenhayn (1861–1922), Chief of the General Staff of the German Army, on the possibility of using lethal chemical weapons.[4] He was also involved in projects to increase the production of ammonia and nitric acid for the manufacture of explosives and fertilisers and the production of the toluene needed to make the high explosive TNT.

Although less well known than Fischer, German chemist and Nobel laureate Heinrich Wieland (1877–1957) similarly carried out extensive research on natural products and contributed his knowledge of chemistry to the German war effort. As with Fischer, his interests in organic chemistry were wide ranging. In 1911, he started to investigate the alkaloid morphine, the analgesic of choice during the First World War, and eventually clarified its structure. For more than 20 years, starting in 1912, he used classical organic chemistry techniques to investigate the chemical structures of the acids in bile, a viscous liquid produced by the liver. He also made an extensive study of fulminic acid. Mercury fulminate, the mercury salt of the acid, detonates spontaneously and violently when subjected to shock or friction, or when heated or hit by a spark. The compound was used as a primer in pistol and rifle cartridges and to fire propellant charges and detonate fuses in rounds of artillery ammunition.

In 1913, Wieland took up a senior lectureship in chemistry at the University of Munich and then, in 1917, was appointed professor of chemistry at the nearby Technical University of Munich. During the last two years of the war, he also worked on chemical warfare projects at the Kaiser Wilhelm Institute for Physical Chemistry and Electrochemistry at Dahlem, Berlin, one

Figure 4.2   Heinrich Wieland (courtesy: Archiv der Max-Planck-Gesellschaft, Berlin-Dahlem).

of the two Kaiser Wilhelm institutes that had been founded in 1911. He developed a procedure for making mustard gas and is also credited with the first synthesis of the organoarsenic compound diphenylaminechlorarsine, a chemical the Germans developed as a chemical warfare agent in the First World War. The compound is also known as Adamsite after American chemist Roger Adams (1889–1971) who synthesised it independently at the University of Illinois in 1918.

Adamsite causes severe irritation to the eyes and mucous membranes of the nose and throat resulting in violent sneezing and coughing followed by bad headaches, acute chest pains, and eventually nausea and vomiting. Chemical warfare agents that induce violent sneezing are known as sternutators. The Americans produced and stockpiled Adamsite as a chemical weapon in World War I with the intention of using it in cloud gas attacks "against an enemy's forward elements to cause temporary casualties."[5] The Germans also manufactured and tested the chemical during the war but it was not used on the battlefield because it readily decomposed when heated.[6]

In December 1928, Wieland received the 1927 Nobel Prize in Chemistry "for his investigations of the constitution of the bile acids and related substances." Compared with Fischer (see Chapter 18) Wieland led a fairly untroubled life. He continued to carry out research in natural products chemistry, publishing scientific papers right through to the early 1940s. Some of his later research focused on toad poisons and curare—a South American Indian arrow poison extracted from the bark of certain species of South American trees. He was happily married, enjoyed family life, took an interest in the arts, and "did not take part in politics," according to one of his biographers.[7]

The Kaiser Wilhelm institute in Dahlem where Wieland worked in the latter part of the war was directed by fellow countryman and future chemistry Nobel laureate Fritz Haber. Haber recruited the team of gas pioneers that carried out the first gas attack on the Allies in April 1915 (see Chapter 12). They included three other future Nobel laureates: Otto Hahn (1879–1968) who won the chemistry prize in 1944, and James Franck (1882–1964) and Gustav Hertz (1887–1975) who shared the 1925 physics prize. Haber's friend Richard Willstätter (1872–1942), director of the other Kaiser Wilhelm institute in Dahlem and joint winner of the 1914 Nobel Prize in Chemistry, also engaged in chemical warfare work, albeit on the design and development of gas masks for the German Army.

Haber not only pioneered chemical warfare for Germany, he also discovered a way of converting the nitrogen in air into a form that could be used for the manufacture of explosives, but that needed the help of Carl Bosch (1874–1940), another German Nobel laureate. Bosch developed Haber's synthesis of ammonia from nitrogen and hydrogen on an industrial scale and was jointly awarded the 1931 Nobel Prize in Chemistry for his work. Germany's Nobel connections to the war effort and the manufacture of explosives did not stop there, however. The Nobel Prize in Chemistry for 1909 was awarded to German chemist Friedrich Wilhelm Ostwald (1853–1932) who discovered the process for converting ammonia into the nitric acid needed to make nitro-explosives such as TNT.

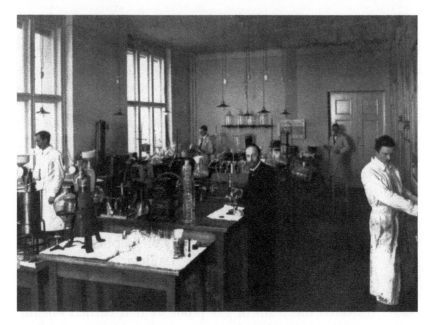

Figure 4.3   Richard Willstätter at the Kaiser Wilhelm Institute for Chemistry, Dahlem, Berlin (courtesy: Archiv der Max-Planck-Gesellschaft, Berlin-Dahlem).

ALLIED POWERS

Two years after Ostwald won the Nobel Prize, Polish-born French physicist Marie Curie (1867–1934) won the Nobel Prize in Chemistry in recognition "of her services to the advancement of chemistry by the discovery of the elements radium and polonium, by the isolation of radium and the study of the nature and compounds of this remarkable element," according to the Nobel citation. She had already been awarded the Nobel Prize in Physics for 1903 jointly with two other French physicists—her husband Pierre Curie (1859–1906) and Antoine Henri Becquerel (1852–1908)—for their research on radioactivity.

In 1914, Marie Curie was in the process of establishing the Radium Institute, now the Curie Institute, in Paris. The outbreak of war put the plans on hold. Curie soon became an enthusiastic advocate of the use of radiography, a technique that was then in its infancy. There was little provision for it in military hospitals at the beginning of the war. "She decided to apply her scientific

Figure 4.4   Pierre and Marie Curie at work in a laboratory (Wellcome Library, London).

talents to training radiologists and radiographers to aid in the defense of the French homeland," observes Roger Macklis, a physician at the Department of Radiation Oncology, Cleveland Clinic Foundation, Ohio.[8]

Assisted by her 17 year old daughter Irène (1897–1956), Marie began to organise X-ray services to assist medical staff locate broken bones, bullets and shrapnel in wounded soldiers.[9] She set up some twenty mobile radiology trucks, later to become known as "*petite Curies*," that carried portable but primitive X-ray machines for use in military field hospitals at the front. The machines used radon as a source of radiation. Radon is a radioactive gaseous chemical element produced by the radio-active decay of radium. At the end of the war, the Radium Institute recruited many of the doctors and scientists who had been trained in radiology and radiography as soon as they were released from active military duty. Her organisational skills and unstoppable drive brought her great acclaim, according to Macklis.

In 1921, she made a six week trip to the United States, arriving in New York on 11 May aboard the White Star Line's ocean liner RMS (Royal Mail Ship) *Olympic*, a sister ship to RMS *Titanic*. She received many honours, tributes and gifts during her visit.

MME CURIE AND HER DAUGHTER IRÈNE, 1925

Figure 4.5   Marie Curie and her daughter Irène (Wellcome Library, London).

On 20 May, for example, Warren Harding (1865–1923), the President of the United States, received her at the White House in Washington, DC, and presented her with a gram of radium, valued at USD 100 000 (equivalent to more than USD 2.3 million in 2014), purchased with funds contributed "by the people of the United States."[10] Four days later, she visited the laboratories of the Welsbach Company, a gas mantle manufacturing firm at Gloucester, New Jersey. At the end of her visit, the firm gave Curie 50 milligrams of mesothorium, one of the many isotopes of radium.

A year after Marie Curie's death in 1934, Irène and her husband Frédéric Joliot-Curie (1900–1958) shared the Nobel Prize in Chemistry for their work on radioactive elements. They had conducted their research at the Radium Institute.

Four years before Marie Curie's visit to the United States, another French wartime scientist and Nobel laureate was also honoured in the country. Victor Grignard (1871–1935), a professor of chemistry at the University of Nancy, had won the Nobel Prize in Chemistry for 1912 for his discovery of the so-called Grignard reagents, a class of organic compounds containing magnesium and a halogen such as chlorine or bromine. The reagents are widely used by chemists to prepare certain types of organic compound. Grignard was awarded the 1912 prize jointly with Paul Sabatier (1854–1941), a chemistry professor at the University of Toulouse. Sabatier had shown that finely-divided nickel and other metals can speed up the reactions of organic chemicals with hydrogen. It was the first time that the Nobel Prize in Chemistry had been shared.

As soon as war broke out, Grignard received his mobilisation orders and served briefly in the French infantry. He was then given the task of finding ways of improving the supplies of the toluene required to make TNT. Following the German chlorine attack in Belgium on 22 April 1915, France launched a chemical warfare research programme. Grignard was called on to head a team of academic chemists, pharmacists, and assistants to investigate the toxic gases and find antidotes for them. They carried out their experiments in a chemistry laboratory at the Sorbonne, a university in Paris. Later in the war, the team studied ways of synthesising phosgene and also developed a precise test for detecting mustard gas.

After the United States entered the war in 1917, Grignard, who was now a major in the French Army, travelled across the Atlantic as a member of a high level scientific commission. "He has come to us at the request of the National Research Council to confer with the Chemistry Committee of the Council and with our War and Navy Departments, and to give us the benefit of the experience which two years of war have brought to the chemical profession in his country," noted an editorial in the September 1917 issue of the *Journal of Industrial and Engineering Chemistry* published by the American Chemical Society (ACS).[11] At its annual meeting held in Cambridge, Massachusetts, the same month, the society recommended that Grignard be elected as an honorary member of the society.[12]

The following month, on the evening of 19 October, the Chemists' Club of New York City similarly conferred honorary membership of the club upon Grignard.[13] The ceremony was conducted during the regular joint meeting of the New York sections of the London-based Society of Chemical Industry, the ACS, and the American Electrochemical Society.

In his address at the meeting, Grignard made clear his antagonism towards Germany and emphasised the importance of science in the war. "In order to save the world from Prussian hegemony, in order to stop the grasping hand of Germany over every liberal thought and generous instinct, we cannot have enough of the support of the United States alongside that of the European Allies," he said. "And this brotherhood of our armies has made indispensable, more particularly between our two countries, the scientific and industrial alliance which is necessary to conduct this war—a war more scientific than anyone ever might have imagined."

The United Kingdom also had its share of Nobel laureates who contributed to the Allied war effort. One of them was Scottish chemist Sir William Ramsay (1852–1916) who won the Nobel Prize for Chemistry for 1904 for his discovery of inert gases. Ramsay was professor of chemistry at University College, Bristol, from 1880 until 1887 and then at University College, London, until his retirement just before the start of the First World War. In 1894, Ramsay and English physicist Lord John Rayleigh (1875–1947), who was professor of natural philosophy at the London-based Royal Institution of Great Britain, announced the

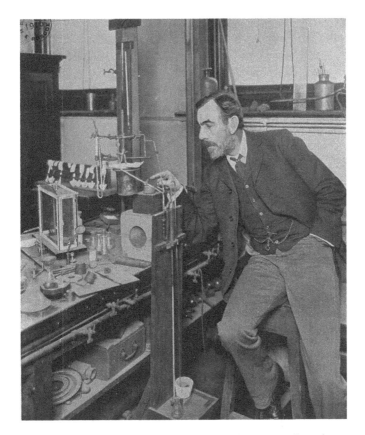

Figure 4.6   William Ramsay (Royal Society of Chemistry Library).

discovery of argon. They had worked independently on the project in their own laboratories but had collaborated closely on the project. Over the next few years, Ramsay and his co-workers isolated other inert gases–helium in 1895 and neon, krypton, and xenon in 1898.

Ramsay was a keen advocate of developing links between academia and industry and was particularly concerned that British manufacturers were not recruiting sufficient numbers of well-trained chemists and exploiting their talents.[14] He was president of Britain's Society of Chemical Industry from 1903 to 1904 and of the Chemical Society from 1907 to 1909.

His research "covered practically every domain of chemistry" noted an Institute of Chemistry obituary published in 1916.[15]

His study of "the decomposition of ammonia by heat" had intrigued Haber and inspired him to pursue his investigations into the synthesis of the gas from its elements (see Chapter 12). In other work, Ramsay carried out experiments on acrolein, also known as propenal or acrylic aldehyde. As the liquid has an unpleasant acrid smell, Ramsay suggested to the British War Office that it might be used as a tear gas on the battlefield. The War Office decided not to use it. Then, early on in the war, Ramsay recommended filling aerial bombs with hydrogen cyanide but once again the War Office rejected his proposal.[16]

Following his retirement in 1912, Ramsay "did little direct chemical research to aid England's war effort," according to one report.[17] But he did contribute in other ways. "With characteristic ardour Ramsay threw himself into every patriotic movement and never relaxed his efforts to rouse and inform the Government and to stimulate the Board of Inventions until overtaken by the dreadful disease which in a few months found the only possible relief in death." The dreadful disease was cancer. He died at the age of 63 on 22 July 1916.

Ramsay's collaborator Lord Rayleigh won the Nobel Prize in Physics in 1904 for his discovery of argon and his investigations of gases. Rayleigh became Chancellor of Cambridge University in 1908. During the First World War, Rayleigh headed an advisory council of eminent scientists that made recommendations to the government on the research that needed to be carried out to enhance the supply of important raw materials, chemicals, and pharmaceutical products that had previously been imported from Germany. He was also president of the government's Committee on Explosives, a position he held from 1896 until 1919. After the war broke out, the committee established a team of chemists that carried out research on explosives such as TNT and cordite.

London-born chemist Cyril Norman Hinshelwood (1897–1967) was just 19 years old when he started to work at His Majesty's Explosives Factory at Queensferry in North Wales during the war. He had just won a scholarship to Oxford University but had to postpone his studies there until after the war. Within two years, he had become deputy chief chemist at the factory. "His remarkable ability was quickly noticed," remarked his obituarist, English chemist Sir Harold Thompson (1908–1983).[18] Thompson

Figure 4.7   Cyril Norman Hinshelwood (Royal Society of Chemistry Library).

had been tutored by Hinshelwood at Oxford and later worked with Haber in Berlin. He was president of the International Union of Pure and Applied Chemistry (IUPAC) from 1973 to 1975 but is more famous in Britain as chairman of the country's Football Association from 1976 to 1981.

The plant at Queensferry had the capacity to produce some 500 tons of TNT and 250 tons of nitrocellulose each week after it opened in December 1915. While working there, Hinshelwood began to investigate the decomposition of solid explosives by measuring the amounts of gas evolved during the process. The work sparked his interest in the mechanism of chemical reactions. His research on reaction mechanisms eventually led to him winning a share of the Nobel Prize in Chemistry for 1956.

English chemist and Nobel laureate Walter Norman Haworth (1883–1950) made a completely different contribution to Britain's war effort. In 1912, Haworth was appointed lecturer in chemistry in the United College of the University of St Andrews,

Scotland. He soon became interested in the chemistry of carbohydrates. At the outbreak of war he abandoned his academic research and arranged for the chemistry laboratories at St Andrews to be used for the production of drugs and other chemicals that Britain now desperately needed.

He resumed his research after the war and in 1920 became professor of organic chemistry at the University of Durham. Five years later he was appointed professor of chemistry the University of Birmingham. By the late 1920s, he had managed to determine the structures of sugars such as lactose, maltose and sucrose. His three-dimensional representations of the structures of carbohydrates became known as "Haworth projections.

In other work, he synthesised and confirmed the structure of vitamin C, which he called ascorbic acid. That was in 1933. Four years later he was awarded the Nobel Prize in Chemistry for his research on carbohydrates and vitamin C jointly with Swiss

Figure 4.8   Walter Norman Haworth (Royal Society of Chemistry Library).

chemist Paul Karrer (1889–1971) who had conducted research on natural products including vitamins A and B_2 and was, incidentally, Wieland's Royal Society obituarist.[7] The initial sample of vitamin C that Haworth worked on had been sent to him by Budapest-born Austro-Hungarian biochemist Albert Szent-Györgyi (1893–1986) who had discovered that Hungarian paprika was a rich source of the vitamin. He gave the name hexuronic acid to the vitamin.

Szent-Györgyi was mobilised as a medical officer when the First World War started and subsequently served in the Austro-Hungarian Army on the Italian and Russian fronts. He was discharged in 1917 after being wounded in action, reportedly by shooting himself in the arm and claiming the injury was caused by enemy fire.[19] He won the Nobel Prize in Physiology or Medicine for 1937 for his discoveries relating to vitamin C and the way biological organisms convert fuel nutrients such as carbohydrates and fats into energy.

REFERENCES

1. J. A. V. Butler and P. Zuman, "Jaroslav Heyrovský", *Biogr. Mems Fell. R. Soc.*, 1967, **13**, 167.
2. L. R. Sherman, Jaroslav Heyrovský (1890–1967), *Chemistry in Britain*, Dec. 1990, p. 1165.
3. G. B. Kauffman, Emil Fischer: His role in Wilhelmian German industry, scientific institutions, and government, *J. Chem. Educ.*, 1984, **61**, 504.
4. D. Charles, *Between Genius and Genocide: the Tragedy of Fritz Haber, Father of Chemical Warfare*, Pimlico, London, 2006, p. 155.
5. A. M. Prentiss, *Chemicals in War: A Treatise on Chemical Warfare*, McGraw-Hill, New York, 1937, p. 275.
6. A. M. Prentiss, *Chemicals in War: A Treatise on Chemical Warfare*, McGraw-Hill, New York, 1937, p. 213.
7. P. Karrer, Heinrich Wieland. 1877–1957, *Biogr. Mems Fell. R. Soc.*, 1958, **4**, 341.
8. R. M. Macklis, Scientist, technologist, proto-feminist, superstar, *Science*, 2002, **295**, 1647.
9. N. Pasachoff, Marie Curie, *Chem. & Eng. News*, 27 June 2011, p. 66.

10. Anon., Madame Curie receives gram of radium and many honors, *Ind. Eng. Chem.*, 1921, **13**, 573.
11. Anon., Greetings to Professor Grignard, *Ind. Eng. Chem.*, 1917, **9**, 636.
12. Anon., Annual meeting American Chemical Society, *Ind. Eng. Chem.*, 1917, **9**, 922.
13. Anon., France and America in scientific union, *Ind. Eng. Chem.*, 1917, **9**, 1142.
14. G. K. Roberts, 'A plea for pure science': the ascendancy of academia in the making of the English chemist, 1841–1914, in *The Making of the Chemist: The Social History of Chemistry in Europe 1789–1914*, ed. D. Knight and H. Kragh, Cambridge University Press, Cambridge, 1998, p. 117.
15. Anon., *Proc. Inst. Chem. GB. Irel.*, 1916, **40**, D033.
16. G. Hartcup, *The War of Invention: Scientific Developments, 1914–18*, Brassey's Defence Publishers, London, 1988, p. 95.
17. G. B. Kauffman and P. M. Priebe, The Emil Fischer-William Ramsay friendship: The tragedy of scientists in war, *J. Chem. Educ.*, 1990, **67**, 93.
18. H. Thompson, Cyril Norman Hinshelwood. 1897–1967, *Biogr. Mems. Fell. R. Soc.*, 1973, **19**, 374.
19. Remembering Albert Szent-Györgyi, 19 March 2014, http://www.history.com/news/remembering-albert-szent-gyorgyi/.

CHAPTER 5

Powering the War

EXPLOITING THE PLANET'S RESOURCES

The chemical industries of the belligerent nations in the First World War drew on and exploited virtually every type of raw material resource on the planet. The ores of the Earth yielded iron for steel production, railways, battleships, and guns; copper for the brass needed to make cartridges and shell cases; the aluminium used to make lightweight alloys for aircraft frames; and tin for the food cans that fed the troops. Minerals such as nitrates found in South America provided the nitrogen and oxygen for explosives, and if these nitrates were not readily available, as was the case in Germany during the war, the nitrogen and oxygen in air could be exploited for the same purpose. Water yielded the hydrogen for airships and natural salt (sodium chloride) the chlorine for chemical warfare and disinfectants.

The industries also relied extensively on plant and animal resources. The acetone needed to make smokeless powders such as cordite was made from wood and other natural sources. Cordite contained two explosives: guncotton (nitrocellulose) made from cotton and nitroglycerine made from glycerine derived from plant and animal fats and oils.

But if there was one raw material on which the war depended more than any other, it was coal. This abundant material had

The Chemists' War, 1914–1918
By Michael Freemantle
© Freemantle, 2015
Published by the Royal Society of Chemistry, www.rsc.org

two primary uses for the war machine and on the home front. They can be summarised in two words: "combustion" and "carbonisation." Combustion of coal raised the steam that powered battleships, merchant vessels, passenger liners, steam locomotives, industrial machinery, pumping and lifting engines in mines, and electricity-generating power stations. Coal was also burnt as a fuel for domestic, commercial and industrial heating, or, as the First World War Song put it, to "Keep the Home Fires Burning." Carbonisation of coal by pyrolysis or destructive distillation produced coke and other products of critical importance, not least coal gas and coal tar.

Most of the Allied and Central Powers' coal production between 1914 and 1918 was geared to the war effort. The quantities produced were immense. For example, Britain's annual production of coal during the war varied from just over 230 million tons to almost 270 million tons.[1] Germany produced almost as much although before the war it also imported coal from England.

A STEAM-POWERED WAR

The British Royal Navy that imposed the blockade of Germany during the war and fought the famous Battle of Jutland on 31 May and 1 June 1916 entered the war in August 1914 with 22 dreadnoughts and 500 cruisers, destroyers, and other types of steam-powered battleships.[2] Altogether, the Allied and Central Powers' navies comprised well over 1200 combat ships at the start of the war, excluding submarines.

Although many of the ships burnt oil as a fuel, others, particularly the larger ones, relied on coal for their power. For example, HMS (His Majesty's Ship) *Dreadnought*, built in Portsmouth and launched in 1906, had the capacity to carry almost 3000 tons of coal to power its steam turbines. The ship's main claim to fame in the war was its ramming and sinking of a German submarine with the loss of all its crew in waters off the north coast of mainland Scotland on 18 March 1915.

Coal was also used to fuel hospital ships, fleets of coastal and ocean-going passenger steamships, as well as the merchant vessels that transported food, minerals, war materials and other vital supplies from around the world to Europe during the war. They inevitably consumed considerable quantities of coal.

Allied fleets faced enormous risks particularly from German U-boats. Over 3300 British merchant ships alone were sunk in the First World War.[3] Even passenger vessels and hospital ships were not immune from attack. On 7 May 1915, a German U-boat fired a torpedo at the British ocean liner RMS (Royal Mail Ship) *Lusitania* sinking it with the loss of 1198 passengers and crew. The ship, owned and operated by Cunard Line, was built at Clydebank, Scotland, and launched in the same year as the *Dreadnought*. It had the capacity to carry 7000 tons of coal.

The White Star Line HMHS (His Majesty's Hospital Ship) *Britannic*, built in Belfast and launched in 1914, suffered a similar fate. The liner, a sister ship to RMS *Titanic* that sank in the Atlantic on 15 April 1912, was powered by steam raised in 29 coal-fired boilers.

The Royal Navy had requisitioned the *Britannic* to serve as a hospital ship in November 1915. Its primary task was to carry sick and wounded troops from Moudros. The port was the Allied base on the Greek island of Lemnos during the ill-fated Gallipoli campaign that took place from April 1915 to January 1916. The ship had a crew of 675 and was equipped with over 2000 berths and 1000 cots for casualties. Almost 500 medical staff, including 100 nurses, cared for the sick and wounded.

In one of its voyages, the ship left Moudros on 3 January 1916 with about 3300 sick and wounded who had embarked from smaller hospital ships. The *Britannic* stopped at Naples to take on coal and water and then finally made port at Southampton on 9 January. The casualties were then put onto ambulance trains and transferred to hospitals around the country.

On 21 November 1916 the ship was steaming at speed in the Aegean Sea when it hit a mine laid by a German U-boat. It sank in less than one hour. Of the 1066 people on board, 1036 survived and 30 lost their lives.

Steam locomotives, like steamships, burnt considerable quantities of coal during the war. A steam engine tender typically had the capacity to carry some five or six tons of the fuel. The railways that criss-crossed Britain not only carried sick and wounded troops and soldiers on leave away from the ports, they also carried the British armed forces and war supplies to the ports. Overall, some 25 million tons of stores, including food,

Figure 5.1   An ambulance train in World War I (Wellcome Library, London).

fodder for horses, coal, and artillery pieces were shipped to France in the war,[4] much of it through Southampton.

When the war started, Southampton was designated No. 1 Military Embarkation Port. By the end of the war, the port had handled 17 186 vessels as part of the war effort.[5] The logistics were formidable. More than eight million British, Colonial and American troops, over 850 000 horses, almost 180 000 vehicles including 2000 aeroplanes, 115 266 guns, eight million tons of stores and over seven million parcels passed through the port between 1914 and 1918. In addition, 1 177 125 sick and wounded along with a steady flow of refugees and prisoners of war disembarked at the port. All this activity relied on coal in one way or another.

France and Belgium were reluctant to employ their steam locomotives near the Western Front. Britain therefore sent their own. In 1916, the Railway Operating Division of the Royal Engineers began to transport scores of locomotives of various types across the English Channel for military use. The division requisitioned the engines from various British railway companies and also placed orders for hundreds more, particularly for heavy goods locomotives. In addition, Britain shipped almost four million tons of coal and some one million tons of railway material to France during the war.[6]

ELECTRICITY GENERATION

From the 1890s onwards, the generation of the electricity used for lighting, cooking, and a variety of other applications in homes, commerce and industry, relied extensively on coal. Near London, two of the earliest coal-fired electricity-generating stations, the Deptford East Power Station and the Kingston Power Station, were commissioned in 1891 and 1893, respectively. By the start of the First World War, scores of such power stations were operating in Britain.

Many of the chemicals and other materials needed for the war effort relied on electricity for their production. Commercial electric arc furnaces that were developed in the years running up to the war generated sufficiently high temperatures to melt scrap iron and steel and other metals for the manufacture of high-quality alloy steels. Silicon steels, for example, were used in the manufacture of munitions, tungsten steels for high-speed machine tools, and nickel steels for the armour plate that protected battleships and tanks.

Construction of First World War aircraft depended on materials generated in electric furnaces and by electrolysis. Chrome alloy steels produced in the furnaces were used to make crank shafts and engine parts. The aluminium used to make alloys for aircraft frames was produced electrolytically by passing an electric current between two electrodes dipped in a molten mixture of alumina (aluminium oxide) and cryolite, an aluminium-containing mineral. Pure aluminium formed at one electrode and sank to the bottom of the vessel containing the molten mixture.

Electrolysis also formed the basis of refining other metals such as the copper that was used to make some of the components of shell fuses, for the manufacture of electrical apparatus and for other purposes central to the war effort. The chloralkali process, another process that relied on electrolysis, similarly played a pivotal role in the war. The chloralkali industry produced huge quantities of electrolytic soda (sodium hydroxide), hydrogen, and chlorine for a multiplicity of applications. Sodium hydroxide, was important for the manufacture of soap, airships were filled with hydrogen, and chlorine was used not only as a war gas but also as a water disinfectant.

PYROLYSIS OF COAL

Many materials, when heated in a closed vessel in the absence of air, decompose to form other substances. The process is known as pyrolysis. For centuries, the process was carried out in Europe to convert bituminous coal into coke. Bituminous coal is a soft coal containing a tarry substance called bitumen. The coal is intermediate in quality between lignite, a poor quality coal, and anthracite, a hard black coal with a relatively high carbon content. Bituminous coal consists of a complicated mixture of chemical substances containing carbon, hydrogen and oxygen along with small amounts of nitrogen, sulfur and some trace elements. Pyrolysis removes the coal's volatile constituents leaving behind a residue consisting mainly of carbon. The residue is called coke.

In the years leading up to the First World War, millions upon millions of tons of coke were produced every year in Britain and other industrialised countries. In 1912, for example, coke production in the United Kingdom amounted to 10.7 million tons. Much of the coke was required for the production of iron in blast furnaces. Coke burns at temperatures approaching 2000 °C in blast furnaces producing carbon monoxide and carbon dioxide. The carbon monoxide reacts with the iron oxides of ores such as haematite yielding molten iron and carbon dioxide which escapes into the atmosphere. The iron was used to manufacture steel for railways, shipping, heavy industry, armaments, and a host of other applications.

The production and use of coke has a long history. Towards the end of the 18th century, dome-shaped firebrick ovens, known as beehive coke ovens, began to be employed for making coke. The ovens had two openings, one through which coal was charged, and another through which coke was withdrawn. By the second half of the 19th century, tens of thousands of beehive ovens were in operation on British coalfields. The Durham coalfield in North East England alone had some 14 000 beehive ovens in 1870.[7]

The flammable volatile constituents of coal produced in beehive ovens were burnt to supply heat for the ovens and the gases and smoke formed by the process vented into the atmospheric causing considerable pollution. The coke alone was preserved and valuable by-products were not recovered. The use of beehive ovens reached a peak in the early 20th century and then gradually declined. The last British beehive oven, at a colliery in County Durham, was closed in 1958.[8]

In the 1880s, by-product recovery coking plants began to be built in the United Kingdom and over the following decades they gradually replaced the beehive non-recovery ovens. By-product coke ovens were designed not only to produce coke as a main product of the destruction distillation of coal but also to recover valuable by-products, principally coal gas, ammonia, and coal tar.

The coal gas manufacturing industry, on the other hand, produced coal gas as the main product, and coke, ammonia, and coal tar as by-products. Coal gas works began to be built in increasing numbers in Britain in the first half of the 19th century, several decades before the advent of by-product coke ovens.

Coal gas contains a mixture of flammable gases, most importantly carbon monoxide, hydrogen, methane, and volatile hydrocarbons such as benzene and toluene. The gas was produced at gas works by heating coal in retorts. The crude gas was then purified, stored in gas holders and finally supplied to users, initially for street lighting. The London-based Gas Light and Coke Company was established in 1812 to manufacture and supply coke and coal gas. The following year the company began to supply gas to light Westminster Bridge and the Houses of Parliament.[9] The first gas works in Paris and Berlin were erected in 1815 and 1826, respectively.[10]

In 1855, German chemist Robert Wilhelm Bunsen (1811–1899) designed a burner that burnt coal gas with a hot non-luminous non-smoky flame. The design relied on mixing the gas with air before it reached the flame. By the end of the century, coal gas (also known as town gas) was being widely used not only for lighting but also as a fuel for cooking and heating.

Both coke and coal gas manufacturing plants produced ammonia and coal tar as by-products. Ammonia gas was processed to form a concentrated ammonium hydroxide solution (called ammoniacal liquor) or a solution of ammonium sulfate. Sulfate was crystallised out of the sulfate solution, dried, and sold as fertilizer. During the First World War, ammonia, which is a form of fixed nitrogen, became critically important in Germany for the manufacture of explosives (see Chapter 9).

Figure 5.2   Robert Wilhelm Bunsen (RSC Library).

The compound had many other uses at the time of the war, notably in the manufacture of dyes for military uniforms and as a coolant in refrigerators. One of the most important was its use in the ammonia-soda process for making the sodium carbonate (soda ash) that was used in vast quantities for manufacturing glass. Soda ash was also used as washing soda, for water treatment, and in many other applications.

Early on in the war, chemists soon realised that the hydrocarbons recovered from coal gas could be important for the war effort. Britain, for example, experienced a shortage of the toluene (also known as toluol) needed to manufacture the high explosive TNT. According to a report in the May 1915 issue of *The Journal of Industrial and Engineering Chemistry*, there was twenty to forty times as much of the hydrocarbon in coal gas made from a ton of coal as there was in the tar produced from the same amount of coal.[11] On the other hand, the removal of toluene and other hydrocarbons such as benzene from coal gas necessarily reduced "the heating value and candle-power of the gas," noted one scientist in an article published in the same journal three years later.[12]

The magnitude of coal consumption and coke production in Britain just before the war can be judged by figures for 1913 presented by a speaker at the Chemical Section of a British Association meeting held in Manchester in September 1915.[13] About 40 million tons of the 189 million tons of coal consumed in the country that year were carbonized—20 million tons in gas works, primarily for the manufacture of town gas, and another 20 million tons in coke ovens for the manufacture of coke for blast furnaces. "Two-thirds of the latter were carbonised in by-product recovery plants; the remainder in the old wasteful beehive ovens," the speaker noted.

COAL TAR AND ITS DISTILLATION

Coal tar, a thick black viscous liquid, is one of the products of the pyrolysis, or more specifically the destructive distillation, of bituminous coal. Each ton of coal yielded some 30 litres of tar during the process. In the early days of coal gas production at the beginning of the 19th century the tar was generally discarded, typically by dumping it into nearby rivers. It "remained for more than a generation from the first introduction of gas-lighting, a nuisance and hardly anything else," observed German chemist George Lunge (1839–1923), a chemistry professor at the Federal Technical University, Zurich, in the first part of his immense work on coal tar and ammonia.[14]

As the century progressed Germany began to find uses for "gas-tar", as Lunge called it, for example, for making waterproof roofing-felt. In Britain, the greatest proportion of coal tar was used for road-making, according to Richard B. Pilcher who was Registrar and Secretary of Britain's Institute of Chemistry during the First World War.[15]

In 1838, British solicitor and engineer John Bethell (1804–1867) took out a patent for a process to preserve timber by impregnating it with coal tar creosote oil. The treatment prevented insect attacks and the decay of the wood. The practice of using tar to treat timber was not new. Tar obtained by the pyrolysis of wood had been used since the 18th century to waterproof the timber used to build the hulls of ships and boats and to seal ropes. Wood tar was mostly produced in heavily forested countries as a by-product of charcoal production.

The heavy creosote oil used by Bethell for his process was obtained by the distillation of coal tar. It was not the first example of coal tar distillation. Scottish chemist Charles Macintosh (1766–1843) used light oil distilled from coal tar as a solvent for the latex used in the manufacture of the waterproof rubber materials for his famous Macintosh raincoats. He patented his invention in 1823 and the first Macintoshes were sold the following year.

Bethell's discovery in 1838, however, demonstrated the value of coal tar distillation and heralded the beginning of the increasingly prosperous coal tar distillation industry. English engineer Isambard Kingdom Brunel (1806–1859) was quick to recognise the importance of Bethell's discovery. Brunel had designed the SS (steamship) *Great Western*, an oak-hulled coal-fired paddle steamer which, in 1838, became the first steamship to cross the Atlantic.

At the time, Brunel was chief engineer of the Great Western Railway that was being constructed between London and Bristol. He realised that Bethel's creosote preservative could not only be employed to protect ships' timbers but also to treat railway sleepers. After taking out a licence to use Bethell's patent, he was instrumental in the construction of a factory in Bristol to distil coal tar obtained from Bristol Gas Works. William Butler, whom Brunel had employed on the railway, was appointed manager of the coal tar distillery and in 1863 bought the business.

"Coal tar creosote eventually became the substance of choice for wood preservation, not least because the huge expansion of the gas industry in the nineteenth century increased its availability and reduced its cost," note British chemists Colin Russell and John Hudson in their book on early railway chemistry.[16]

The distillation of coal became big business following Bethell's discovery and Brunel's adoption of the process. Butler's company not only produced creosote oil for preserving railways sleepers, it also produced pitch, the residual product of coal tar distillation. Pitch comprised between 50% and 65% of the coal tar distilled at the tar works. The company used the pitch to bind otherwise unusable coal particles and dust into fuel briquettes.

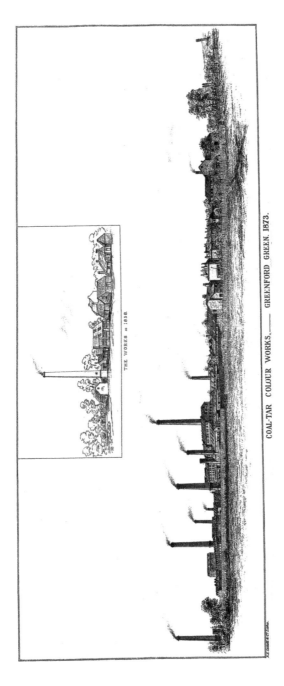

Figure 5.3 Coal tar colour works at Greenford, near London, in 1858 (top) and 1873. The works belonged to William Henry Perkin Snr and his brother Thomas Dix Perkin (Wellcome Library, London).

In an article on coal tar pitch published in September 1916, American chemical engineer John Morris Weiss (1885–1963) noted that the pitch was used in the United States for roofing purposes, for waterproofing bridges and tunnels, for the construction of pavements, for paints and protective coatings, and for briquetting fine coal and coke particles. He confessed, however, that "very little is known of the actual nature of the compounds making up coal-tar pitches."[17]

T. Howard Butler, managing director of the William Butler Company, made a similar point in a book on distillation published in 1922.[18] "The composition of coal tar ... varies so enormously, according to its mode of production, that it is useless to attempt to give any definite analysis," he noted. The same was true for other types of tar, for example wood tar, although that was rarely distilled.

The distillation of coal tar yielded several fractions that boiled in distinct temperature ranges. The most volatile fraction, Butler observed, contained naphtha, which is a flammable mixture of benzene, toluene and other light hydrocarbons, and ammoniacal liquor. The fraction boiled between 80 °C and 140 °C. The other fractions described by Butler were a light oil fraction with a boiling point range of 140 °C to 200 °C; a middle oil fraction, also known as the carbolic oil fraction, as it contained phenol (carbolic acid), that boiled between 200 °C and 240 °C; a creosote oil fraction (240 °C to 280 °C); and a coloured fraction variously known as the green oil, yellow oil, or anthracene oil fraction with a boiling point greater than 280 °C.

By the time of Bethell's discovery, chemists in Europe were already beginning to identify organic chemicals in coal tar and isolate them from the tar, albeit on a laboratory scale. Naphthalene was extracted in 1819 and within 15 years anthracene, aniline, and phenol had also been recovered. Then, in 1845, German chemist August Wilhelm von Hofmann (see Chapter 2) reported finding benzene in coal tar. Five years later, one of his students, English chemist Charles Mansfield (1819–1855), showed that benzene could be prepared in a pure state on a large scale by the fractional distillation of coal tar. He died at the age of 35 from injuries sustained when a still containing coal tar naphtha overflowed and caught fire in his laboratory.

Within another ten years more than 200 organic chemicals had been isolated from coal tar. By the turn of the century, benzole, a crude form of coal tar benzene, was being used as a fuel for motor cars. In the First World War it was also employed as a shell propellant.

A SERENDIPITOUS DISCOVERY

In the first half of the 19th century, coal tar produced by the coke and gas industries was principally used to build roads, preserve wood, and make raincoats, but supply of the tar far exceeded demand. That was to change with the development of industrial processes for distilling coal tar, the identification of various chemical components of coal tar, and in particular a serendipitous discovery in the late 1850s. The discovery opened up a new market for coal tar, had a major impact on the German economy before the First World War, and ultimately underpinned and sustained the war effort on the land, in the air, and at sea of both the Allies and the Central Powers.

In 1856, William Henry Perkin, an English chemist and assistant to Hofmann, fortuitously discovered how to make a synthetic dye. While carrying out research on chemicals obtained from coal tar in a laboratory in his family home in London, he produced a purple dye from toluidine, an impurity in aniline, a nitrogen-containing organic chemical derived from coal tar.

The discovery revealed for the first time that an organic chemical originating from coal tar could be converted into a commercial and potentially hugely profitable product. The following year, Perkin set up a small factory to manufacture the synthetic dye, known as mauve or mauveine, and subsequently made a fortune from the process. He gave much of the money away to charity.

Perkin's discovery revolutionised fashion, transformed the textiles industry, and attracted immense interest in the commercial application of synthetic organic chemistry. Over the following decades, chemists discovered ways of converting coal tar chemicals into numerous dyes and other useful chemical products that could be manufactured inexpensively on an industrial scale. Within 50 years, some two thousand artificial

Figure 5.4   William Henry Perkin Snr (RSC Library).

colours had been synthesised and used to dye fabrics, leather, paper, wood, and other materials. Perkin's discovery also led indirectly to major advances in medicine, perfumery, food, explosives, and photography, observes British author Simon Garfield in his biography of the chemist.[19] Germany in particular rapidly began to exploit these developments in synthetic organic chemistry. Coal tar benzene, for example, was readily converted into nitrobenzene which in turn could be used to manufacture the aniline needed for the production of aniline dyes. Phenol was obtained in relatively small amounts directly by the distillation of coal tar and in much larger amounts from benzene by its reaction first with sulfuric acid and then with caustic soda (sodium hydroxide).

By the start of World War I, Germany had become the world leader in the production and export of synthetic organic chemical products, notably dyes and pharmaceuticals. On the other hand, Germany imported large quantities of coal tar and its more important chemical components from Britain in the years leading up to the war in order to manufacture these commercially valuable synthetic products. It also imported pitch from Britain to make fuel briquettes and other products.

The war put a stop to this trade between Britain and Germany. As dyes were needed for military and naval uniforms and medicines for the care of the sick and wounded, the British Board of Trade had to take prompt action to replace the synthetic organic chemical products previously imported from Germany. Among its initiatives, it signed an agreement with dye manufacturers in Switzerland, a neutral country in the war, to export dyes synthesised from coal tar chemicals supplied by Britain. It also pressed British chemical companies to step up their production of dyes, medicines, and other chemical products required for the war effort.

British companies such as William Butler supplied toluene distilled from coal tar for the manufacture of TNT and other coal tar chemicals essential for the war effort. Phenol (carbolic acid) extracted from coal tar or prepared from coal tar benzene was perhaps one of the most important organic chemicals in the war. It was needed to make Lyddite—a high explosive used to fill shells. The pale yellow explosive, otherwise known as trinitrophenol or picric acid, was produced by the nitration of phenol. In France, the explosive was known as melinite.

Carbolic acid and picric acid were both used in antiseptic preparations for the care of the sick and wounded in the war. In addition, carbolic acid was widely employed as a disinfectant in the trenches and casualty clearing stations and as an ingredient of carbolic soaps such as Wright's Coal Tar Soap. Aspirin, an antipyretic and analgesic also known by its chemical name acetylsalicylic acid, was synthesised from phenol. It was the world's best-selling drug before the war and was administered extensively throughout the conflict (Figure 5.5).

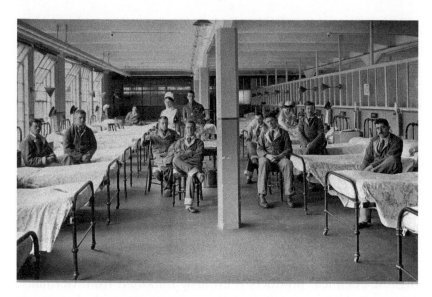

Figure 5.5   Wounded soldiers in a ward at a military hospital during the war. Many of the anaesthetics, antiseptics, and drugs used in hospitals derived from coal tar (The RAMC Muniment Collection in the care of the Wellcome Library).

The tar-distilling industry in Britain produced coal tar on a huge scale over the course of the war to satisfy the huge demand for fuel, explosives, dyes, medicines, and other coal tar products required for the war effort. Returns from the industry collated by the Ministry of Munitions during the war show that Britain distilled an average of almost 1.4 million tons of coal tar annually in the years 1914 to 1916 and more than 1.6 million tons in both 1917 and 1918.[20]

CALCIUM CARBIDE AND ACETYLENE

On 2 May1892, Canadian chemist Thomas L. Willson (1860–1915) heated a mixture of lime (calcium oxide) and coal in an electric furnace and showed that the resulting material produced a flammable gas when thrown into water. The same process occurred when carbon in the form of coke and lime were heated together in the furnace. Chemists subsequently showed that the material was calcium carbide and the gas formed by its reaction with water was acetylene.

The discovery sparked numerous applications that proved invaluable in the First World War, not least for the production of explosives. In 1915, for example, the Canadian Electro Products Company was established to convert calcium carbide into chemicals required for the British war effort. One such chemical was acetone which was needed to make the propellant cordite. The Canadian process had three stages. The first produced acetaldehyde by the catalysed hydration of acetylene. The acetaldehyde was then oxidised, once again with a catalyst, resulting in the formation of ethanoic acid (commonly known as acetic acid). Finally, the acid was passed over a hot catalyst to make the acetone.

Calcium carbide absorbs atmospheric nitrogen at high temperatures to produce calcium cyanamide. The cyanamide reacts with water to yield ammonia, a compound that Germany exploited in the war for the manufacture of the nitric acid needed to make explosives such as TNT. Cyanamide also proved to be important in agriculture as it decomposes in soil to form urea and ammonium carbonate, both of which are nitrogenous fertilisers. The country produced tens of thousands of tons of cyanamide in the war, most of it for the production of explosives (see Chapter 9).

The acetylene produced by the hydrolysis of calcium carbide burns in air with a bright flame. The process was exploited in acetylene gas lamps, also known as carbide lamps. A typical lamp worked by allowing water in an upper reservoir to drip slowly into a chamber containing granules of the carbide. The acetylene generated by the reaction was then ignited with a match or lighter. The lamps were used to illuminate coal mines, buildings, marine buoys, and lighthouse beacons, as well as for motor car and bicycle headlights.

Acetylene burns in oxygen to give flame temperatures of 3000 °C or more, temperatures that are hot enough for welding. The first commercial oxyacetylene welding apparatus was developed in France in 1901. In the Great War, the apparatus was used to weld aircraft parts together. British official war artist Christopher Nevinson (1869–1946) depicted the operation in the image "Acetylene Welder," one of six lithographs on paper dated 1917 that he created during the war. They formed part of series of 66 lithographs entitled "Building Aircraft."

The acetylene produced from calcium carbide was also employed as the starting material for the syntheses of organic compounds that were important for the war. Butadiene is an example.[21] In 1910, Russian chemist Sergei Vasiljevich Lebedev (1874–1934) showed that synthetic rubber could be made by the polymerisation of butadiene. Rubber was needed for the manufacture of hoses, belts, gaskets, tyres and other key components of First World War machinery. Because of the British Royal Navy blockade, Germany was starved of natural rubber and so it turned to Lebedev's process to make synthetic rubber. The country developed the process and managed to manufacture some 2500 tons of a poor-quality synthetic rubber known as methyl rubber by the polymerisation of dimethyl butadiene.

REFERENCES

1. J. Ellis and M. Cox, *The World War I Databook*, Aurum Press, London, 2001, p. 285.
2. J. Ellis and M. Cox, *The World War I Databook*, Aurum Press, London, 2001, p. 251.
3. G. Bridger, *The Great War Handbook*, Pen & Sword, Barnsley, 2009, p. 48.
4. G. Bridger, *The Great War Handbook*, Pen & Sword, Barnsley, 2009, p. 191.
5. Rance, p. 142.
6. G. Bridger, *The Great War Handbook*, Pen & Sword, Barnsley, 2009, p. 192.
7. S. H. Beaver, Coke manufacture in Great Britain: A study in industrial geography, *Trans. Pap. (Inst. Br. Geogr.)*, 1951, **17**, 133.
8. D. G. Edwards, By-product coking plants in Britain: An outline history, 2 October 2013, http://www.coke-oven-managers.org/PDFs/edwards.pdf.
9. R. B. Pilcher, *What Industry Owes to Chemical Science*, The Scientific Book Club, London, 1947, p. 228.
10. G. Lunge, *Coal-tar and Ammonia*, Part I, 4th Edn, Gurney and Jackson, London, 1909, p. 8.
11. M. L. Hamlin, Recovery of toluene from gas, *Ind. Eng. Chem.*, 1915, **7**, 438.
12. R. S. McBride, Toluol recovery and standards for gas quality, *Ind. Eng. Chem.*, 1918, **10**, 111.

13. W. A. Tilden, *Chemical Discovery and Invention in the Twentieth Century*, 3rd Edn, George Routledge and Sons, London, 1919, p. 303.
14. G. Lunge, *Coal-tar and Ammonia*, Part I, 4th Edn, Gurney and Jackson, London, 1909, p. 9.
15. R. B. Pilcher, *What Industry Owes to Chemical Science*, The Scientific Book Club, London, 1947, p. 235.
16. C. A. Russell and, J. A. Hudson, *Early Railway Chemistry and its Legacy*, RSC Publishing, Cambridge, 2012, p. 65.
17. J. M. Weiss, Coal-tar pitch, *Ind. Eng. Chem.*, 1916, **8**, 841.
18. T. Howard Butler, Fractional distillation in the coal tar industry, in *Distillation Principles and Processes*, ed. S. Young, Macmillan, London, 1922, p. 361.
19. S. Garfield, *Mauve: How One Man Invented a Colour that Changed the World*, W. W. Norton, New York, 2001, p. 8.
20. T. Howard Butler, Fractional distillation in the coal tar industry, in *Distillation Principles and Processes*, ed. S. Young, Macmillan, London, 1922, p. 365.
21. American Chemical Society National Historic Chemical Landmarks, *Discovery of the Commercial Processes for Making Calcium Carbide and Acetylene*, American Chemical Society, Washington, DC, 1998.

CHAPTER 6

The Chemistry of a Single Firearm Cartridge

NESTS OF SNIPERS

"The smallest of big game animals did not present so small a mark as the German head, so that sniping became the highest and most difficult of all forms of rifle shooting." So wrote British sniping expert Hesketh Hesketh-Prichard (1876–1922) in his memoirs of the First World War.[1] The multi-talented Hesketh-Prichard was a big-game hunter who also played cricket as a fast bowler for Hampshire and other teams before 1914. During the war, he was a press officer for the British War Office. While on the Western Front, he became concerned at the level of casualties inflicted by German snipers and by the poor standard of British sniping. "Suffice it to say that in early 1915 we lost eighteen men in a single battalion in a single day to enemy snipers," he observed.[2] A battalion comprised some 1000 troops. Hesketh-Prichard's observations led him to introduce training courses in 1915 to improve the marksmanship of British snipers.

With the advent of static trench warfare on the Western Front in October 1914, each battalion in the front line began to establish nests of snipers at regular intervals along its sector of the line—the intervals ranging from 500 metres early on in the war to 40 or 50 metres as the war progressed. "Each British battalion

The Chemists' War, 1914–1918
By Michael Freemantle
© Freemantle, 2015
Published by the Royal Society of Chemistry, www.rsc.org

had at least four specialist snipers," notes British author Tony Ashworth in a book about trench warfare.[3] British snipers worked in pairs, one with a telescope who acted as an observer to spot the enemy and the other with a rifle, the standard issue being the Short Magazine Lee Enfield Mark III repeating rifle. The SMLE rifle, as it was called, was shorter than earlier versions of the rifle. It was loaded with a box magazine containing ten rounds of ammunition and could fire bullets at more than twice the speed of sound.

The standard German service rifle, the Mauser Gewehr 98, was loaded with a five-round magazine and when fitted with a telescopic sight was highly accurate. Like the SMLE rifle, it fired bullets at more than the speed of sound. So, if troops in the trenches heard the loud crack of a rifle being fired and the whistle, whir or whine of the bullet, they knew they need not duck out of the way. The bullet had already passed them by.

MUNITIONS FUSIL BRITANNIQUE
Ammunition for British rifle (Lee Enfield)
Munition für das britische Gewehr (Lee Enfield)
303 British (7,7mm)

Figure 6.1 Ammunition for the Lee Enfield rifle on display at The Somme Trench Museum, Albert, France (author's own, 2010, Musée Somme, 1916).

But of course, bullets fired from magazine-fed high velocity snipers' rifles did hit their targets and if they did not kill their victims outright, they caused horrendous injuries. A single bullet could become embedded in the flesh of an arm or leg, smash a bone into small fragments, lodge in the stomach, rip out an eye, pierce the skull, or pass right through the body. A lot depended on the type of bullet, its velocity on hitting the victim, and the point of entry, or in some cases the points of entry.

A single bullet could make four or more wounds, noted Sir John Bland-Sutton (1855–1936) in a lecture on bullet wounds that he presented at the Middlesex Hospital, London, in 1914.[4] Bland-Sutton, a surgeon at the hospital and a major in the Royal Army Medical Corps (RAMC), observed that in some cases, a soldier might be injured by a bullet passing through both legs creating two entry and two exit wounds. In one instance, a man in a British Army cavalry regiment in the surgeon's care received five wounds from one bullet. "It passed through both buttocks, leaving four holes, and in passing across the interspace between the two buttocks it chipped a piece of skin off the posterior aspect of the scrotum," Bland-Sutton reported. "The soldier recovered."

"If a bullet strikes a man's pocket it will carry in fragments of a pipe, or a piece of a tobacco pouch, beside pieces of cloth, leather, coins, *etc.*," he continued. "A bullet moving with a high velocity will traverse a limb and perforate a large bone like the femur and cause very little disturbance; or it may catch a nerve, an artery or a vein and produce instantly unmistakable signs"— the signs being lots of blood and excruciating pain.

THE ISSUE OF RIFLES AND OTHER FIREARMS

Rifles were issued not just to snipers but to almost every soldier in the war. They were particularly important weapons for the infantry when attacking across open ground but of much less use for close combat fighting in the trenches. The rifle strength (the number of officers and men) in a British infantry division— consisting of three brigades of four battalions—was about 12 000. A German infantry division also had a strength of some 12 000 rifles whereas each Russian infantry division consisted of around 16 000 rifles.

Figure 6.2 A Christmas postcard showing two soldiers in a trench, one writing a letter. Rifles were issued to almost every soldier in the war (The RAMC Muniment Collection in the care of the Wellcome Library, London).

At the beginning of the war, many armies experienced critical shortages of weapons and ammunition and consequently issued their troops with obsolete rifles that had been phased out prior to the war. The various industries that supported the war effort therefore had to move into overdrive towards the end of 1914 to supply the needs of the military.

In Russia, for example, the engineering industry supplied about 25% of its production to the military before the war.[5] That increased to almost 70% by 1916. When the war broke out, the country had a deficiency of about 350 000 rifles.[6] The shortage resulted in many recruits arriving at the front without rifles. Some went into battle unarmed and were instructed to arm themselves with rifles removed from their dead comrades. Others were sent back from the front. In 1916, the Russian munitions industry produced well over 1.3 million rifles, almost five times the output in 1914, and by June 1917, the country had a stock of over 4.5 million rifles.[7]

Apart from rifles, other types of firearm were also used extensively in the First World War. They included handguns such as the German Luger P08 and American Colt M1911 semi-automatic pistols that were fed with ammunition from detachable box magazines containing seven and eight rounds respectively. Germany manufactured some two million Luger pistols during the war. The Webley Mark VI revolver with a revolving cylinder that held six rounds of ammunition was introduced by the British as the standard issue service pistol in May 1915. The French Berthier Mousqueton carbine was widely used by a number of the smaller Allied armies in the war. They included the Greek Military Forces who purchased them from France after Greece entered the war on the side of the Allies in May 1917. Carbines are essentially similar to rifles except that they have shorter barrels.

A POWERFUL DEFENSIVE WEAPON

One of the most devastating of all firearms used in the war was the machine gun. Infantry or cavalry attacking across open ground were easily raked with machine gun fire and liable to be mown down like blades of grass under a lawn mower. One of the battle scenes in the film *War Horse* depicts British cavalry chasing Germany soldiers over open ground and then charging into a wood only to be faced by the withering fire of German machine guns. The film was produced and directed by the American Steven Spielberg and based on the book by British author Michael Morpurgo.[8] The scene illustrates the power of the machine gun and its importance as a defensive weapon in the war.

The first day of the Battle of the Somme provided a notable if not notorious example of how the machine gun played a critical role in the war. At 7.30 a.m. on 1 July 1916, the first waves of British and French infantry emerged from their trenches and began to advance across No Man's Land towards German lines. The attack followed an intensive artillery bombardment over several days that was intended to destroy the German defences. The French and British south of the River Somme and the town of Albert made some notable gains. But in the north, the bombardment proved ineffective and the German trenches and barbed wire defences opposite the British lines remained intact. The British troops in northern sectors walked at a steady pace through a hail of German machine gun bullets and artillery shells and were killed or wounded in their thousands. By the end of the day, of the 120 000 British troops who took part in the attack, 19 240 were killed and another 38 230 were wounded. It was the greatest loss of life in a single day in the history of British warfare.

"Fighting machine gun bullets with the breasts of gallant men," was how future British Prime Minister Winston Churchill (1874–1965) famously described such attempts to break through the German lines.[9] But it was not just the British infantry who suffered. On the same day, the fighting claimed some 7000 French casualties. Around 4000 German troops were also killed or wounded and 2000 captured.

In another example, troops landing on narrow beaches at Anzac Cove in the Dardanelles, a narrow strait in northwest Turkey, had to face the heavy machine gun and rifle fire of Turkish infantry positioned on top of the cliffs of the Gallipoli peninsula. The landing took place on 25 April 1915 on the first day of the ill-fated Gallipoli campaign against the Ottoman army.

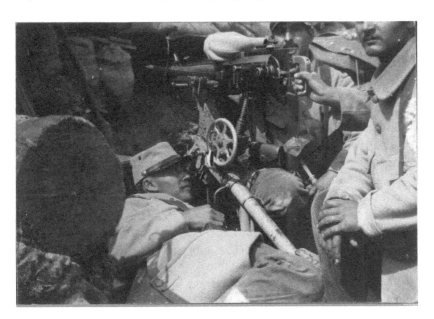

Figure 6.3   Three men at a machine gun post, possibly in France, 1917 (The RAMC Muniment Collection in the care of the Wellcome Library, London).

The cove is named after the Australian and New Zealand Army Corps who took part in the amphibious invasion.

Machine guns were used not just by the infantry in the front line. Both British and German tanks were equipped with machine guns in the world's first-ever tank-against-tank battle that took place at Villers-Bretonneux, France, on 24 April 1918 during the German Spring Offensive (see Chapter 15).[10] Tanks armed with both cannons and machine guns were known as "males" and those with machine guns only as "females."

THE DEVELOPMENT OF THE MACHINE GUN

Like so much technology used in the First World War, the machine gun was not new to the war. The origin of the machine gun can be traced back to the early 18th century when English inventor James Puckle (1667–1724) developed a revolver-like gun with a single barrel and a hand-operated rotating cylinder that could fire nine shots a minute, which is three times the rate that muzzle-loading muskets of the time could be fired. Puckle patented the gun in 1718. In 1862, American inventor Richard Gatling (1818–1903) patented a rapid-fire gun with six parallel barrels that were rotated manually. For ammunition, the Gatling gun used the metal cartridges that had been invented and developed in the 1840s and 1850s. The gun was first used in combat, albeit to a limited extent, in the American Civil War (1861–1865). Later versions of the gun had ten barrels and were able to fire five hundred rounds of ammunition per minute.

Whereas the Puckle and Gatling guns required manual operation, the Maxim machine gun was fully automatic. When a gun is fired it kicks or springs back, a process known as recoil. The Maxim gun exploited recoil to eject the spent cartridge after firing and then load another cartridge. Sir Hiram Stevens Maxim (1840–1916), who was born in the United States but emigrated to England in 1881, patented the gun in 1883 and introduced a prototype the following year. He became a British citizen in 1900 and was knighted the following year.

All the European armies in the First World War used machine guns based on Maxim's design. Models such as the British Vickers machine guns were known as heavy machine guns as they could weigh around 45 kg when fully loaded.[11] Because of

Figure 6.4   Vickers machine gun on display at The Somme Trench Museum, Albert, France (author's own, 2010, Musée Somme, 1916).

their weight, they were primarily used for defence. The Vickers gun was water cooled to prevent overheating. A team of three or more men mounted the gun on a tripod and fed ammunition into the breech of the gun using a fabric belt holding 250 cartridges. The gun fired some 500 rounds per minute, but although reliable and efficient to operate, it was unwieldy.

The Lewis gun, a light machine-gun weighing around 13 kg when loaded, could be carried by a single infantryman in an assault. The American-designed gun had a mechanism that exploited both recoil and the pressure of gas generated by the propellant when it was fired to eject a spent cartridge and insert the next round. The air-cooled gun was loaded with a circular magazine holding 47 cartridges and was capable of firing over 500 rounds per minute. The British introduced the Lewis gun for service for the infantry in 1915 and subsequently used it to arm tanks and aircraft.

During the Spring Offensive in 1918, the Germans equipped infantry assault troops with a submachine gun known as the Bergmann MP 18. It was the first time that such an automatic weapon had been employed in warfare. Whereas machine guns such as the Vickers and Lewis guns were designed to fire rifle-calibre ammunition, the MP 18 fired shorter pistol-calibre cart-ridges fed from a 32-round detachable drum magazine. Submachine guns such as the MP 18 had shorter barrels and were more manageable than other types of machine gun in use at the time. Even so, they only played a minor role in the war and had little impact on the outcome.

MACHINE GUN SUPPLIES

Machine guns of one sort of another were manufactured and deployed in increasingly large numbers as the war progressed. Russia, for example, had a little more than 4 000 machine guns in 1914.[7] That worked out at just one per infantry battalion. By June 1917, stock of the guns had risen to well over 18 000. The use of machine guns on such a scale inevitably required millions upon millions of rounds of ammunition. Overall, the country's munitions factories produced some four billion cartridges for its machine guns and rifles between 1914 and 1917.[12]

As with the Russian army, the British Expeditionary Force ex-perienced a severe shortage of machine guns at the start of the war. Even in 1916, a British infantry battalion was fortunate if it had four Lewis light machine guns. Following Churchill's ap-pointment as Minister of Munitions in July 1917, the supplies rose steadily. By 1918 each battalion was equipped with around thirty Lewis guns. With a similar abundance of other firearms, artillery weapons, shells, aircraft and tanks, "the British could now fight a rich man's war, with unlimited supplies of matériel," comments British military historian Gary Sheffield in one of his books on the First World War.[13]

A Maxim gun such as the Vickers could fire up to 10 000 rounds per hour. There are reports of some British companies firing almost one million rounds in 24 hours.[14] At the end of the war, Britain alone had a stock of more than 325 million of its standard issue machine gun and rifle cartridges left unused.[15]

THE CHEMISTRY OF A SINGLE ROUND

Machine gun and rifle fire accounted for some 39% of the 2.9 million British casualties in the First World War.[16] Every single one of the bullets that killed, maimed or missed a soldier was a product of chemistry. Likewise, every single round of firearm ammunition fired by a pistol, a sniper's rifle, or a Vickers machine gun, relied on chemical powders and chemical processes to propel that bullet out of the barrel of the gun towards its victim.

A round of firearm ammunition, also known as a cartridge, is essentially a compact chemical system. The cartridges used in the war typically comprised four chemical components: a percussion cap, a propellant, a bullet and a case that contained the other three components. When a soldier pulled the trigger of a gun, he released a firing pin that struck the percussion cap. The explosive inside the cap, known as the primary explosive or primer, exploded with a flash that ignited the propellant. Combustion of the propellant rapidly generated gases at pressures sufficiently high to propel the bullet out of the cartridge case and gun barrel at high velocity.

Whereas a firearm cartridge is a complete system, the cartridge used by the artillery to fire shells is just one part of a round of artillery ammunition. An artillery cartridge contains the propellant that projects the shell. In the First World War, the cartridge was either loaded separately into the gun, or it was attached to the base of the shell. The other parts of the shells used in the war were: the priming device that ignited the propellant; the shell itself which contained a high explosive, a chemical warfare agent, or shrapnel balls; and a fuse attached to the nose or base of the shell.

THE BRITISH .303 CARTRIDGE

The standard service firearm cartridge issued by the British Army for its Lee Enfield rifles, Vickers machine guns, and Lewis light machine guns was the ".303", so named because the calibre, that is the internal diameter of the gun barrel, was 0.303 inches (equal to 7.7 millimetres). The cartridge's percussion cap was made of copper. It contained a pellet of percussion powder consisting of an explosive mixture of chemicals, also known as "the cap composition" or "primer composition." The cap was prepared by carefully pressing the mixture into the copper with an anvil. It was then often covered with a disc of tinfoil so that it could be readily stored in small magazines.[17]

The most important of the chemicals used in cap compositions during the war and throughout the 19th century and
early 20th century were mercury fulminate, antimony trisulfide,
and potassium chlorate.[18] Mercury fulminate, also known as
fulminate of mercury or mercuric cyanate, is a primary explosive
that detonates spontaneously and violently when subjected to a
shock or friction, hit by a spark, or heated. The heat and flame
caused by the explosion in the cap when struck by a firing pin
ignited the antimony trisulfide, a combustible compound. The
oxygen needed for the combustion was supplied by potassium
chlorate, a widely used oxidising agent.

Percussion powders and the chemicals they contained were so
dangerous they were "usually mixed by hand on a glass-top table
by a workman wearing rubber gloves and working alone in a
small building remote from others," noted American organic
chemist and explosives expert Tenney L. Davis (1890–1949) in his
book *"The Chemistry of Powder and Explosives"* published in
1943.[19]

The percussion cap was inserted into a space in the centre of
the base of the cartridge. For this reason this type of cartridge
was known as a centrefire cartridge. When the firing pin struck
the cap, it crushed and detonated the primary explosive powder
creating a flash that passed through two fire holes in the base
and ignited the propellant.

Early versions of the .303 British cartridge employed gunpowder, also known as black powder, as the propellant. Gunpowder is typically a mixture of 75% by weight of potassium
nitrate, also known as saltpetre, 15% carbon in the form of
charcoal, and 10% sulfur. Ignition of the powder sets off a fairly
complex series of chemical reactions leading to the formation of
nitrogen, carbon monoxide, carbon dioxide and other gases.[20]
The process also produces numerous solid particles of other
chemicals and consequently a lot of smoke. Powders such as
ballistite and cordite, both of which contain nitroglycerine and
nitrocellulose, were subsequently employed as the propellant in
the cartridges as they produced far less smoke than gunpowder
when fired. They were known as "smokeless powders."

The bullet in the .303 cartridge was pointed and consisted
of a cupronickel alloy envelope made of about 80% copper and
20% nickel that sheathed a metallic core composed of two

segments.[21] The front segment was either pure aluminium or an alloy of 90% aluminium and 10% zinc. The rear portion was an alloy of 98% lead and 2% antimony which hardened the lead. The design enabled the bullet to move about from side to side during its flight in a motion known as yawing. The movement inflicted greater injury when the bullet hit a victim.

The three components of the .303 cartridge, that is the primer, the propellant and the bullet, were enclosed in a brass case. Brass is an alloy of copper and zinc. Fabrication of the complete cartridge consequently relied on the production and availability of a variety of chemical materials including four alloys (brass, cupronickel, aluminium–nickel and lead–antimony) manufactured from six chemical elements (copper, zinc, nickel, aluminium, lead, and antimony) and five chemical compounds (mercury fulminate, potassium chlorate, antimony trisulfide, nitroglycerine, and nitrocellulose).

The manufacture of all these materials required a significant input of chemistry. Mercury fulminate, the detonating compound in the primer composition, was prepared by treating a solution of mercury in concentrated nitric acid with ethanol. Nitrocellulose (also known as guncotton) and the nitroglycerine in the cordite propellant were manufactured by the nitration of cotton and glycerine respectively using concentrated nitric and sulfuric acids. The glycerine was produced during the manufacture of soap from fats and oils (see Chapter 8). The starting materials for the production of the potassium chlorate in the primer composition were potash (potassium carbonate) and salt (sodium chloride). Antimony trisulfide, another component of the composition, occurs naturally as the black crystalline mineral stibnite.

Copper—the metal used to make the percussion cap, the cupronickel alloy bullet sheath, and the brass case—was extracted from copper-containing ores by chemical processes and refined electrochemically. The other metals used to make the bullet and cartridge case—zinc for the brass and the front portion of the bullet, nickel for the cupronickel alloy sheath, and the aluminium and lead for the bullet were similarly extracted from ores. Antimony, a component of the alloy in the rear portion of the bullet, is not strictly a metal. It is one of several chemical elements known as metalloids that exhibit some of the characteristics of

metals and some of those of non-metals. Silicon is another met-
alloid. Antimony is produced industrially from stibnite in a process that removes the sulfur from the mineral.

OTHER CARTRIDGES

The manufacture of other types of cartridge used in the First
World War was similarly underpinned by chemistry and the
availability of metal-containing minerals and other natural
resources. The French, for example, produced hundreds of millions of 8 mm Lebel Balle D cartridges for its Lebel rifles and air-
cooled Hotchkiss machine guns. The cartridges employed *Poudre
B*, a smokeless powder consisting of nitrocellulose stabilized
with diphenylamine, as the propellant.

The bullets in these cartridges produced "small and neat"
wounds, according to a report published in June 1915 in the
British Medical Journal.[22] "The French bullet, as a rule, caused
little destruction, even when it penetrated bones or vital organs,
and the prognosis for wounds inflicted by the French was usually
good, while it was much worse for wounds inflicted by the
British," the report claimed. The French bullets were pointed
and made of "a solid mass of bronze" containing 90% copper,
noted another report published in December 1914 in the same
journal.[23] Bronze is an alloy of copper and tin.

The German service cartridge, the Mauser S Patrone, like the
British .303 and French Lebel cartridges, was loaded with a
sharp-nosed bullet known as a "spitzer," a word derived from
the German term *Spitzgeschoss*, meaning "pointed bullet." The
German bullet consisted of a solid lead core covered by a steel
envelope coated with cupronickel.[24] The bullet had a sharper
nose than the British and French bullets and inflicted nasty
wounds, especially as the envelope could peel off and the core
split into minute fragments if the bullet bounced off a stone or
some other hard object before it hit its victim.[4]

REFERENCES

1. H. Hesketh-Prichard, *Sniping in France: With Notes on the
Scientific Training of Scouts, Observers, and Snipers*, Hutchinson, London, 1920, p. 26.

2. H. Hesketh-Prichard, *Sniping in France: With Notes on the Scientific Training of Scouts, Observers, and Snipers*, Hutchinson, London, 1920, p. 2.
3. T. Ashworth, *Trench Warfare 1914–1918: The Live and Let Live System*, Pan, London, 2000, p. 57.
4. J. Bland-Sutton, Value of radiography in the diagnosis of bullet wounds, *Br. Med. J.*, 1914, 2, 953.
5. P. Gatrell, *Russia's First World War: A Social and Economic History*, Pearson Education, Harlow, 2005, p. 119.
6. G. Jukes, *The First World War: The Eastern Front 1914–1918*, Osprey, Oxford, 2002, p. 17.
7. P. Gatrell, *Russia's First World War: A Social and Economic History*, Pearson Education, Harlow, 2005, p. 119, p. 232.
8. M. Morpurgo, *War Horse*, Egmont, London, 2007.
9. W. Davis, *Into the Silence: The Great War, Mallory, and the Conquest of Everest*, Bodley Head, London, 2011, p. 132.
10. D. R. Higgins, *Mark IV vs A7V: Villers-Bretonneux 1918*, Osprey, Oxford, 2012, p. 10 and p. 16.
11. M. Pegler, *British Tommy: 1914–1918*, Osprey, Oxford, 1996, p. 18.
12. P. Gatrell, *Russia's First World War: A Social and Economic History*, Pearson Education, Harlow, 2005, p. 119, p. 120.
13. G. Sheffield, *The Chief: Douglas Haig and the British Army*, Aurum Press, London, 2011, p. 293.
14. M. Pegler, British Tommy, in *The Chief: Douglas Haig and the British Army*, ed. G. Sheffield, Aurum Press, London, 2011, p. 107.
15. G. Bridger, *The Great War Handbook*, Pen & Sword, Barnsley, 2009, p. 120.
16. R. M. MacLeod, Chemistry for King and Kaiser, *Determinants in the Evolution of the European Chemical Industry, 1900–1939*, ed. A. S. Travis, H. G. Schröter, E. Homburg and P. J. T. Morris, Kluwer Academic Publishers, Dordrecht, The Netherlands, 1998, p. 26.
17. T. L. Davis, *The Chemistry of Powder and Explosives*, Vol. II, John Wiley & Sons, New York, 1943, p. 455.
18. E. M. Chamot, *The Microscopy of Small Arms Primers*, Cornell Publications, Ithaca, New York, 1922, p. 9.
19. T. L. Davis, *The Chemistry of Powder and Explosives, Vol. II*, John Wiley & Sons, New York, 1943, p. 454.

20. M. S. Russell, *The Chemistry of Fireworks*, Royal Society of Chemistry, Cambridge, 2000, p. 6.
21. War Office, *Treatise on Ammunition*, 10th Edn, The Naval & Military Press, Uckfield, East Sussex, and The Imperial War Museum, London, 1915, p. 531.
22. Anon., German views on British bullets, *Br. Med. J.*, 1915, **1**, 1023.
23. Anon., German, French, and British bullets, *Br. Med. J.*, 1914, **2**, 990.
24. War Office, British and German Small Arm Ammunition, *Br. Med. J.*, 1914, **2**, 895.

The Acetone Crisis

THE 1915 SHORTAGES

One crisis can lead to another. That was certainly the case in 1915 when the British Expeditionary Force (BEF) in France faced a severe shortage of artillery shells that became known as the "shell crisis of 1915." The crisis, which led to the so-called "shell scandal," was resolved by the end of the year, but not before it had precipitated another but less well known crisis: the propellant or acetone crisis.

Propellants were needed to fire the shells. The standard propellant used by the British Army and the Royal Navy during the First World War was a powder known as cordite. The manufacture of the powder relied on supplies of a number of chemicals, not least acetone, an organic solvent also known as propanone. Acetone was also needed to produce other materials essential for the British war effort. By 1915 acetone was in short supply, the shortage being so severe that it might well have imperilled the ability of both the army and navy to function effectively. Like the shell crisis, the acetone crisis was to have far-reaching political consequences.

The shell scandal broke on 14 May 1915 when BEF Commander-in-Chief Sir John French (1852–1925) asserted that the Battle of Neuve Chapelle had failed because of the lack of

The Chemists' War, 1914–1918
By Michael Freemantle
© Freemantle, 2015
Published by the Royal Society of Chemistry, www.rsc.org

shells. "The want of an unlimited supply of high explosives was a fatal bar to our success," he told Colonel Charles à Court Repington (1858–1925), a British Army officer and a war correspondent for British national newspaper *The Times*.

Neuve Chapelle is a small village in northern France, half way between the cities of Béthune and Lille. The British offensive, which lasted from 10–13 March 1915, aimed to break through German lines at the village, drive on and seize the nearby Aubers Ridge, and finally attack Lille some 15 miles away. Following an intense artillery bombardment guided in part by reconnaissance planes of the Royal Flying Corps, the infantry rapidly breached the German lines and captured the village. The advance then ground to a halt and troops were unable to press their advantage. They abandoned their attempt to capture Aubers Ridge. Eight weeks later, on 9 May, the British once again attacked the ridge but failed to take it. Because of limited supplies, Sir John French had to ration the number of shells fired by British heavy guns. Furthermore, 90% of the shells were shrapnel shells that proved to be almost completely ineffective at cutting through the heavily fortified German defences. The attack was called off the following day.

At the same time, in Turkey, the Allies had just started fighting Ottoman forces in an unsuccessful attempt to secure the Gallipoli peninsula on the northern bank of the Dardanelles strait. The failure combined with the shell crisis led to the fall of the Liberal government led by Prime Minister Herbert Asquith (1852–1928). Asquith formed a coalition government on 25 May, created a new department, the Ministry of Munitions, and appointed David Lloyd George (1863–1945) as its minister. Lloyd George rapidly established a national system of munitions factories to manufacture shells, TNT and other high explosives, propellants such as cordite, small-arms ammunition and other matériel for the British war effort. By the start of 1916, the ministry had set up 73 such factories on new sites in Britain and by the end of the war the number stood at 218.[1]

But these factories required raw materials to make the chemicals needed to manufacture the munitions. The production of cordite, for example, depended on secure supplies of nitroglycerine, guncotton (a highly-nitrated version of nitrocellulose), nitric acid, petroleum jelly and acetone. The glycerine

needed to make nitroglycerine was a by-product of soap manufacture which in turn required a supply of animal fats and plant
oils (see Chapter 8). Guncotton was produced by the nitration of
cellulose, a constituent of cotton and other plants. According the
Treatise on Ammunition published by the British War Office in
1915, cotton waste was used for this purpose.[2] The nitric acid
used to nitrate the glycerine and cellulose was manufactured
from Chile saltpetre, a nitrate mineral imported from Chile and
Peru (see Chapter 11).

The supply of acetone for cordite manufacture was a major
problem. At the start of the First World War, Britain manufactured
little of the solvent itself but instead relied on imports from the
United States to meet most of its needs.[3] By the beginning of 1915
stocks of the chemical were rapidly diminishing and the country
was having difficulty in satisfying the ever-increasing requirements of the national munitions factories which desperately
needed the solvent to make sufficient quantities of cordite for the
army and navy. Within 12 months, the country was in crisis.

A VERSATILE SOLVENT

Acetone is a colourless, volatile and highly flammable liquid with
a distinctive ethereal odour. The liquid is one of the most ver-
satile of all organic solvents. It is miscible with water, most other
organic solvents, and most oils. It is also a good solvent for fats,
waxes, and a wide range of solid inorganic and organic compounds. These properties make it an ideal cleaning and drying
agent for laboratory equipment. In the home, it is the familiar
sweet-smelling component of nail polish remover and superglue.

Nobody knows how or when acetone was discovered. The al-
chemists called the chemical "spirit of Saturn" because it could
be obtained from "salt of Saturn."[4] The salt is now known as
lead acetate (or by its more modern chemical name: lead(II)
ethanoate). However, it was not until the early decades of the
19th century that chemists began to investigate systematically the
structure, physical properties and chemical reactions of acetone.

By the start of the First World War, it was well known that the
cellulose in wood pulp could be converted into cellulose acetate
by its reaction with acetic anhydride and furthermore that certain forms of cellulose acetate dissolved in acetone. A few years

earlier, two Swiss chemists, Camille Dreyfus (1878–1956) and his younger brother Henri (1882–1944), had shown that cellulose acetate–acetone solutions could be used to make non-flammable acetate films for use in photography and also non-flammable acetate lacquers that could be applied as fire-resistant dopes to waterproof and stiffen the canvas wings of aircraft. The cellulose acetates were safe alternatives to celluloid films, which consisted mainly of highly-flammable nitrocellulose, and to the nitrocellulose aircraft dopes that were liable to cause the aircraft to burst into flames if a bullet hit one of its wings.

The First World War started less than eleven years after the two American aviation pioneers, the Wright brothers, Orville (1871–1948) and Wilbur (1867–1912), made their epic first flights in a heavier-than-air petrol-powered aircraft, the *Wright Flyer*, in North Carolina. In the decade leading up to the war, heavier-than-air flying machines began to fly higher and higher reaching heights of up to 5000 feet by 1912. The accidental mortality rate among pilots also fell markedly as the safety features of the aircraft improved. By 1914, Britain had already established a Royal Flying Corp, which was part of the British Army, and the Royal Navy Air Service. As the war progressed, the design, production, and deployment of military and naval aircraft in Britain, France, Germany and other belligerent powers advanced rapidly. The aircraft were used for a variety of purposes including photographic reconnaissance, bombing, and aerial combat.

The British War Office was keen to use the non-flammable cellulose acetate dope for its aircraft and so, in November 1914, commissioned the Dreyfus brothers to produce the material. The brothers came to England and by early 1916 had set up a factory to make the lacquer at Spondon in Derbyshire, but production of the lacquer relied on acetone which, because of its scarcity, was much in demand at the time.

Acetone was also used in the manufacture of tetryl during the war. The high explosive, also known as "composition exploding" or by chemical names such as "trinitrophenylmethylnitramine" or "nitramine," was used in large quantities as a booster explosive in fuses for shells and in some types of high-explosive shells. The explosive was prepared by the nitration of the organic compound dimethylaniline with sulfuric acid and nitric acid.

The crude product was purified by crystallisation from either acetone or benzene.

The greatest need for acetone in Britain, however, was for the manufacture of the propellant cordite. Before the war, the Royal Gunpowder Factory at Waltham Abbey in Essex and several commercial factories had manufactured the propellant, mainly for the Royal Navy. The process at Waltham Abbey, for example, involved running nitroglycerine, an oily liquid, from tanks into lead burettes.[5] A measured volume of the liquid was then run from the burettes onto dried guncotton that had been weighed into rubber-lined canvas bags. The two ingredients were mixed to form a paste which was sieved and loaded into an iron box known as the incorporating machine. An operator then sprayed acetone onto the paste. Finally, petroleum jelly was added to the paste and the resultant dough extruded into cords using a hydraulic press.

Cordite was used as a propellant during the war not only by Britain but also by countries such as Australia, Canada, Japan, and New Zealand. It was not the only propellant to be used by the Allies. France used *Poudre B*, (an abbreviation of *Poudre blanche* meaning white powder) which consisted of two types of nitro-cellulose: guncotton and collodion. Guncotton is more highly nitrated than collodion and insoluble in a mixture of diethyl ether and ethanol whereas collodion is soluble in the mixture. To make the powder, the two materials were kneaded together with a mixture of the two solvents. The antioxidant diphenylamine was then added to stabilise the propellant.

Germany, Austria, Italy and other countries used yet another type of propellant during the First World War. Known as ballistite, it consisted of a mixture of collodion and nitroglycerine in equal proportions. A small amount of the organic chemical camphor was added to stabilise the mixture.

Cordite, *Poudre B*, and ballistite are all known as smokeless powders because they produce far less smoke than gunpowder when used to fire guns. Gunpowder, also known as black powder, consists of a mixture of saltpetre (potassium nitrate), charcoal and sulfur, and had been used as a propellant ever since the invention of the gun in the early 14th century.

Poudre B was the first of the smokeless powders to be introduced. It was initially prepared by French chemist Paul Vieille

(1854–1934) in 1884. Swedish chemist and industrialist Alfred Nobel (1833–1896) patented ballistite three years later and then, in 1889, English chemist and explosives expert Frederick Abel (1827–1902) and Scottish chemist and physicist James Dewar (1842–1923) jointly patented cordite. Dewar is more famous for inventing the vacuum flask, otherwise known as the Dewar flask or thermos flask, in 1892.

Ballistite and cordite are similar in several ways, the major difference being the nature of the nitrocellulose in the mixture and the type of solvent used to produce the propellant. Had Britain opted to use ballistite as its staple propellant for army

Figure 7.1   James Dewar (Royal Society of Chemistry Library).

and navy guns in the First World War, then it would not have faced the acetone crisis of 1915 and 1916, but the country had always preferred cordite, ever since its introduction in 1889.

ACETONE FROM WOOD

Wood was a source of organic chemicals long before the advent of coal tar chemicals in the late 19th century (see Chapter 5) and petrochemicals in the early 20th century. The organic solvent turpentine, for example, was first produced from pine trees in the 16th century. Distillation of the trees' oily resin yielded not only turpentine, but a solid residue of rosin which was used in the manufacture of inks, paints, varnishes and a multitude of other products.

In 1799, French civil engineer Philippe le Bon (1767–1804) showed that wood gas could be obtained by the carbonisation of wood in the absence of air. The gas, which is a mixture of methane, hydrogen, carbon monoxide, carbon dioxide and

Figure 7.2 Wood was an importance source of acetone and other organic chemicals (author's own, 2009).

nitrogen, was used for city lighting in England before the arrival of coal gas (see Chapter 5).

Other wood products of commercial importance before and during the First World War included:

- Natural organic dyes of various colours such as the yellow dyes extracted from the osage orange trees that grew in large numbers in Oklahoma, Texas and elsewhere in the United States;
- Cellulose for making explosives, celluloid photographic films, and cellulose acetate dopes (see above);
- Tannins for tanning animal skins into leather;
- Wood tar creosotes used for wood preservation.

One of the most valuable products was wood alcohol, otherwise known as methanol or methyl alcohol. By the start of World War I, the alcohol was used all over the world and new fields were being discovered for its use all the time, observed Solomon Acree, professor of the chemistry of forest products at the University of Wisconsin–Madison, in a paper on wood chemistry that he presented at the 31st meeting of the American Chemical Society held in Seattle in 1915.[6] "The 10 000 000 gallons produced in this country are used for making celluloid and similar products, dyestuffs, denatured alcohol [methylated spirits], photographic films, formaldehyde, artificial leather, varnishes, shellacs, artificial rubber, and other substances too numerous to mention, in whose manufacture and use hundreds of millions of dollars and large numbers of men are employed," he said. "European countries use nearly as much wood alcohol as the United States but chiefly for denaturing ethyl alcohol, for the manufacture of formaldehyde, for the methyl group of aniline colors, and for other purely chemical purposes."

The wood alcohol was produced by the destructive distillation of wood under airtight conditions. If wood is heated in a plentiful supply of air it burns forming steam, carbon dioxide and eventually wood ash. When oxygen is excluded or partially excluded, the wood distils forming wood gas; a dark liquor known as pyroligneous acid or wood vinegar; and charcoal, which is mainly carbon. The proportions of each of these components depend on the type of wood that is distilled and the way it is distilled.

The composition of each component also varies. According to the Food and Agriculture Organization of the United Nations, one ton (1000 kg) of air dry northern hemisphere deciduous hardwood such as European beech yields a pyroligneous liquor containing 250 kg of tars, 50 kg of acetic acid, 16 kg of methanol and 8 kg of acetone and methyl acetone (a mixture of acetone, methyl acetate and methanol).

Until the early 20th century, acetone "was obtained solely as one of the ultimate products of the distillation of wood," noted British chemist Richard B. Pilcher in his book on chemistry and industry published in 1947.[7] However, the relatively low yields of acetone produced by the distillation were insufficient to satisfy the growing demands for the solvent. Acetic acid, a liquid that had once been known as "burntwood acid,"[8] was present in pyroligneous liquor in far greater quantities and proved to be a much more plentiful source of acetone.

The acetone was produced from the acid by first adding lime, also known as calcium oxide, to neutralise the acid. The lime was obtained by heating limestone or other rocks containing calcium carbonate. After neutralisation, the liquor was distilled and the distillate evaporated allowing solid calcium acetate, or acetate of lime as it was sometimes called, to crystallise. Distillation of the dry solid yielded crude acetone which was then refined by fractional distillation.

In June 1910, British Conservative politician Charles Bathurst (1867–1958) observed in a parliamentary debate in the House of Commons that there was a shortage of acetone in Britain.[9] He suggested that "in the event of war with one of the chief manufacturing countries" the government should establish, "either in the Forest of Dean or the New Forest, a factory for the manufacture of acetone in sufficient quantities to meet all national requirements." The Forest of Dean, in Gloucestershire, England, and the New Forest in southern England both had substantial areas of woodland amounting to hundreds of square kilometres. He noted that Germany was one of the countries with well-developed woodland industries. The following year, in another debate, he asked the government to "consider the advisability of purchasing acetate of lime from local manufacturers for conversion into acetone."[10]

In 1913, the British government set up a small factory in the Forest of Dean but it took around 100 tons of wood to produce just one ton of acetone, and so the output was low. When the First World War started, Britain had a mere 3000 tons or so of the solvent available for military use and other needs.[11] Much of it was imported from the United States. By May 1915 the country had opened two more factories to produce acetone by the destructive distillation of wood but it still was not enough.

Britain was now desperate to find a further source of acetone in order to manufacture the quantities of cordite required by the British Army and Royal Navy. Fortunately another source was available although it took a while for the British government to realise it.

In the years before the war, scientists in Paris and Manchester had shown that it was possible to convert the starch in potatoes and grain into acetone and other organic chemicals by fermentation. One of the scientists, a chemistry lecturer at the University of Manchester, subsequently developed the process on an industrial scale. Although well known in the world of science for his fermentation process, he is more famous as a statesman. His name was Chaim Weizmann.

THE FATHER OF INDUSTRIAL FERMENTATION

While carrying out research on fermentation of maize at Manchester in the years before the war, Weizmann discovered a strain of bacteria that converted the carbohydrates in the grain into acetone and butanol (also known as butyl alcohol). The name of the organism was *Bacillus Clostridium acetobutylicum.* It later became known as the "Weizmann organism."

Weizmann was born on 27 November 1874 in Motol, a small town in Belarus which then formed part of the Pale of Settlement, a poor region of the Russian Empire where Jews were allowed to live. He was the third of 12 children. His father, Ozer, was a hard-working timber merchant, or *"transportierer"* as Weizmann called him in his autobiography *Trial and Error.*[12] "He cut and hauled the timber and got it floated down to Danzig," (a Polish city on the Baltic coast now known as Gdańsk) he noted.

Jews were not encouraged to enrol in Russian universities and so, at the age of 18, the young Weizmann moved to Germany

Figure 7.3   Chaim Weizmann (courtesy: Weizmann Institute of Science).

where he studied chemistry. In 1897 he enrolled at Fribourg University, Switzerland, where he carried out research on dye-stuffs. He obtained his doctorate there in 1899 with the top rating of *summa cum laude*. The achievement brought him a teaching appointment as *"Privatdozent"* at the University of Geneva. By then he had already become an ardent member of the Zionist movement that aimed to establish a Jewish homeland in Palestine. "Early in 1903 I was hard at work both on chemistry and Zionism," he observed.[13]

While in Geneva, he worked under German chemistry professor Carl Gräbe (1841–1927) who specialised in dye chemistry. In 1904, Weizmann decided to move to England partly because he thought that Britain would be able to help establish a national homeland for the Jews in Palestine. He ended up in Manchester speaking little English but with a letter of introduction from

Gräbe to William Henry Perkin, Jr, who was professor of organic chemistry at Victoria University of Manchester and a fluent German speaker. Perkin's father, William Henry Perkin, Snr, had famously synthesised mauve, the first synthetic coal-tar dye, in 1856. The discovery gave birth to the synthetic dyes industry and led to the mass production of the synthetic organic chemicals that were to have an immense impact on the First World War (see Chapter 5).

Perkin Jr was impressed that Weizmann had also worked on dyes and gave him a warm welcome. He arranged for Weizmann to have the use of a university laboratory which, however, turned out to be a dingy grimy basement room "covered with many layers of dust and soot," according to Weizmann.[14] Perkin was a consultant for the Clayton Aniline Company, a dye-manufacturing firm that had been founded by French chemist Charles Dreyfus (1884–1935) in 1876. Dreyfus asked Perkin to look into a possible synthesis of camphor, a naturally-occurring compound that was used not only to plasticise nitrocellulose in the manufacture of celluloid but also had a number of medicinal and other applications. Perkin passed the investigation to Weizmann who worked on the project part-time.

Dreyfus was an influential figure in Manchester at the time. He was chairman of the Manchester Zionist Society, chairman of the local Conservative Association and a member of Manchester City Council. In January1906, he introduced Weizmann to Arthur Balfour (1848–1930), a British Conservative politician who was a Member of Parliament for the Manchester East constituency and British Prime Minister from July 1902 to 1905. It was the first of several meetings.

In 1910, Weizmann became a naturalised British subject and in the same year unsuccessfully applied to become a Fellow of the Royal Society.[15] However, two other events that year sparked Weizmann's interest in the fermentation process that became so important for the production of acetone in World War I.

First, two biologists at the Pasteur Institute in Paris, the Frenchman Auguste Fernbach (1860–1939) and the Latvian Moïse Schoen (1884–1938), showed that it was possible to obtain both butanol and acetone by fermenting the starch in potatoes using what became known as Fernback and Schoen's culture. The production of butanol by fermentation was not new. It was first reported in 1861 by French chemist Louis Pasteur (1822–1895) who is regarded as the father of modern bacteriology. However, it was not until 1905 that the production of acetone by fermentation was first reported.[16]

Second, the British company Strange and Graham launched a project to produce synthetic rubber from either butanol or the related compound isoamyl alcohol (also known as 3-methyl-1-butanol). The idea was to convert one of these alcohols into isoprene or butadiene which could then be polymerised to make synthetic rubber. With the growing demand for rubber, prices were increasing and "there was a clamour for an artificial product," Weizmann remarked in his autobiography.[17] The company therefore decided to explore the possibility of producing the alcohols on an industrial scale by fermentation using Fernback and Schoen's culture. The company recruited Fernbach, Schoen, Perkin Jr, and Weizmann to collaborate on the project.

According to Weizmann, Manchester University lacked the facilities to conduct research in biochemistry and bacteriology and so he "began to pay frequent visits to the Pasteur Institute in Paris." Back in Manchester, he trained himself to become a microbiologist by reading scientific texts and carrying out microbiology experiments in the laboratory.

In 1912, Weizmann severed his connection with Strange and Graham but continued his research on fermentation at Manchester. He started to look for a bacterial fermentation culture that would produce high yields of isoamyl alcohol. He eventually found a bacterium that produced large quantities of a liquid that smelled like the alcohol. Distillation, however, revealed that the liquid was a mixture of acetone and butanol. He isolated the bacterium, the so-called Weizmann organism. The bacterium not only produced higher yields of acetone and butanol than Fernbach and Schoen's culture, it could also be used to ferment the starch in grains, notably maize which contains around 70% by weight of the carbohydrate.

The following year, Strange and Graham began producing acetone and butanol in England from potatoes using Fernback and Schoen's culture. When the war started, the British government contracted the company to supply acetone but its process was inefficient and it was only able to produce about 440 kg of the solvent each week.[16] Together with imported acetone and the acetone produced by the destructive distillation of wood, it was not enough to make all the cordite needed for the British war effort.

In December 1914, C. P. Scott (1846–1932), the editor of the *Manchester Guardian* (now *The Guardian*) newspaper, introduced Weizmann to Lloyd George who at the time was the British Chancellor of the Exchequer. Lloyd George expressed interest not only in Weizmann's research but also in his views on Palestine as a potential homeland for Jews.

Weizmann was keen to publish his discovery of the new organism in a scientific paper but with the outbreak of the war he soon realised that his process might be useful for the British war effort. He therefore decided to patent his discovery and subsequently became chemical adviser to the Ministry of Munitions and the Admiralty.

In March 1916, Weizmann was introduced to Winston Churchill who was then First Lord of the Admiralty. He was "brisk, fascinating, charming and energetic," noted Weizmann in his autobiography.[18] Churchill was impressed by the process which Weizmann had shown could potentially produce around 12 tons of acetone from 100 tons of maize. He asked Weizmann if he could make 30 000 tons of acetone. Weizmann answered: "So far I have succeeded in making a few hundred cubic centimetres of acetone at a time by the fermentation process."

Even so, Weizmann agreed to develop the process and soon started to supervise the construction of a pilot plant at a gin distillery in London. After resigning from his post at Manchester University he took up an appointment as superintendent of the Admiralty laboratories at the Lister Institute in London. Within months, he was able to scale up the process to produce 500 kg batches of acetone with "consistently satisfactory results." He also trained a team of chemists to develop the process at factories and other distilleries in London and Scotland.

One of the factories was at King's Lynn in the east of England. The factory had originally produced oil cake, the solid left over when foods such as apples or olives are pressed to extract juice or oil. The cake was typically used as animal feed. In 1915, the factory was adapted to produce acetone from the starch in

Figure 7.4 Laboratory at the Lister Institute, London (Wellcome Library, London).

potatoes and then, in March 1916, it was nationalised and used to make acetone from maize using the Weizmann process.

Weizmann sent one of his team, a chemist named Alfred Appleyard (1888–1949), to develop the process and run the operation at the factory. Appleyard had been a member of staff at the Rothamsted Experimental Station, an agricultural research institution in Harpenden, a town in Hertfordshire, England. Earlier in 1916, the British had introduced conscription into the armed forces for single men aged between 18 and 41 years, but as Appleyard was a chemist he was exempted and instead sent to work as a research assistant to Weizmann in the Admiralty laboratories at the Lister Institute. He did not run the fermentation process at King's Lynn for long, however. The following year the British government selected him to go to India to set up a new acetone factory for the Indian government.[19]

To boost output of acetone in Britain even more, the Admiralty asked Strange and Graham to adopt the Weizmann process and in addition constructed an acetone plant with fermentation vessels at the Royal Navy Cordite Factory at Holton Heath,

Dorset. By 1917, Weizmann's process was producing some 3000 tons of acetone a year from both maize and rice at the Holton Heath plant and other sites in Britain.

Maize, rice, and the other cereals used as sources of the starch for the fermentation process were also needed for food. Large quantities of these grains were imported from the United States into Britain but with Germany's resumption of unrestricted submarine warfare in 1917 and the growing threat of a U-boat blockade Weizmann began to look for alternative sources of starch. His team carried out experiments on a variety of possible substitutes at the King's Lynn factory. He tested wheat which was grown in Britain. However, the cereal was not plentiful at the time. About 80% of the bread consumed in the country was made from imported wheat. Furthermore, yields of acetone from wheat were not high. Weizmann also searched for non-edible sources of starch that could be converted into acetone and soon discovered that horse chestnuts, or conkers as they are widely known, could be used.

The Ministry of Munitions quickly endorsed the idea and encouraged children to contribute to the war effort by gathering the horse chestnuts. The government did not want the Germans to get to know the reason for collecting the chestnuts and therefore did not tell the public that they were to be used to produce acetone for the manufacture of cordite. In the event, the children collected more conkers than could be transported to the factory. Many of the conkers were left to rot at railway stations and barely 3000 tons arrived at the plant in King's Lynn. Although the factory began to stockpile horse chestnuts in the autumn of 1917, it was not until April 1918 that the factory managed to produce any acetone from them. Production stopped within a few months. The scheme had not been a great success.

The focus of acetone production by the Weizmann process subsequently switched to Canada and the United States. In Toronto, Herbert Speakman, professor of biological chemistry at the University of Toronto and one of Weizmann's former students, had already managed to adapt a distillery to produce significant quantities of the solvent. The plant operated from August 1916 until November 1918. Following its entry into the war in 1917, the Americans built a similar plant for acetone production at Terre Haute, a city in Indiana. It started operations in May 1918 and like the Canadian plant continued producing the solvent until the end of the war.

In the patent of his fermentation process, Weizmann refers to the production of acetone and "alcohols" from starch in cereals.[20] He specifies butanol, or butyl alcohol as he called it, but does not mention any other alcohol. However, it is now well established that one of the alcohols produced by his process is ethanol, or ethyl alcohol as it is also called. For this reason the process is known as ABE fermentation. It produces approximately 30% acetone, 60% butanol and 10% ethanol.

As a result of his work on fermentation during the war, Weizmann became known as the "father of industrial fermentation." He was the first person to demonstrate that an organic chemical other than ethanol could be produced on an industrial scale by fermentation.

His autobiography suggests that the Balfour Declaration of 1917 was a reward for his services to the country. The British government, largely influenced by Lloyd George who was Prime Minister at the time, issued the declaration in November 1917. It recommended the establishment of a national home for the Jewish people in Palestine.

After the war, Weizmann's attention swung away from science to politics. He was elected unopposed as president of the World Zionist Organisation in July 1919 and in the years that followed he worked and travelled tirelessly to promote the movement. The State of Israel was eventually established in Palestine in May 1948. In February 1949, Weizmann became the first president of the country but by the end of the year he had become physically weak and nearly blind.[21] He was re-elected as president in November 1951 taking the oath from his sick-bed. He died in November the following year at the age of 77.

His obituary in the *Journal of the Chemical Society* notes that his fermentation process was "but a further step on a road opened by another Manchester scholar, Arthur Harden, with his studies on the biochemistry of yeast fermentation."[22] Harden (1865–1940), a biochemist, was a lecturer at Manchester for nine years until 1897 when he was appointed as chemist to the Jenner Institute of Preventive Medicine. The institute subsequently changed its name to the Lister Institute. By the time Weizmann arrived at the institute in 1916, Harden was head of its biochemical department. Interestingly, Weizmann does not mention Harden in his autobiography even though both men had taught at Manchester University, both carried out research on fermentation, and both were employed at the Lister Institute at the same time. Harden was knighted in 1926 and in 1929 was jointly awarded the Nobel Prize in Chemistry with Swedish biochemist Hans von Euler-Chelpin (1873–1964) "for their investigations on the fermentation of sugar and fermentative enzymes," according to the citation.

Weizmann's obituarist in the journal was German-born chemist Ernst David Bergmann (1903–1975). Bergmann notes that Weizmann's research was interrupted by his political responsibilities until 1931 when "his longing for scientific work became so overwhelming that he established a private laboratory" in London. Bergmann joined him there in 1933 after Adolf Hitler (1889–1945) became chancellor of Germany. "David Bergman" was one of "Germany's leading scientists," Weizmann wrote.[23] Bergmann left a year later to work at the Daniel Sieff Research Institute at Rehovot in Palestine. Weizmann had been instrumental in setting up the institute and also conducted research in its laboratories.

In 1939, the year that the Second World War broke out, Weizmann was appointed scientific advisor to the Ministry of Supply in Britain and had a laboratory put at his disposal in London. According to Bergmann, Weizmann then resumed his research on "various aspects of the acetone–butyl alcohol fermentation and the utilisation of its products." He also carried out investigations on the synthesis of isoprene and the production of synthetic rubber. Weizmann's scientific work ended for good with the end of the war.

Four years later, the institute at Rehovot was renamed the Weizmann Institute of Science. Bergmann was subsequently appointed the first chairman of the Israel Atomic Energy Commission and went on to play a leading role in the development of the Israeli nuclear programme.

ACETONE FROM ETHANOL

Weizmann was not the only chemist in Britain to work on the production of acetone during the First World War. Another was

Figure 7.5   Nevil Sidgwick (Royal Society of Chemistry Library).

English chemist Nevil Sidgwick (1873–1952) who is most famous
in the world of chemistry for his contributions to the theory of
chemical bonding and valency. He wrote several books on
chemistry, including the highly successful *Organic Chemistry of
Nitrogen* which was published in 1910 and subsequently re-
printed and revised many times.[24]

Sidgwick was born in Oxford and graduated at Oxford University with First Class honours degrees in natural sciences in
1895 and in "Greats" (Roman and Greek history and philosophy)
in 1897. He then went to Germany to pursue the study of
chemistry, "as was fashionable" in those days, remarked fellow
Oxford chemist Leslie Sutton (1906–1992) in his obituary of
Sidgwick.[25] Sidgwick obtained a doctorate at the University of
Tübingen in southern Germany and returned to Oxford in 1901.

When World War I broke out he was a demonstrator in
chemistry at Oxford University. He volunteered as an unpaid
consultant to the Department of Explosive Supplies of the

Ministry of Munitions. According to English chemist Henry Tizard (1885–1959) who had spent some time at the ministry during the war, Sidgwick "was clearly unfitted" for any normal form of military service.[26] At Oxford, Sidgwick worked with Perkin Jr and other chemists on the production of acetone and other projects of importance to the British war effort including the manufacture of phenol from benzene, a coal tar chemical, and the preparation of mustard gas.

Perkin Jr, an organic chemist, had moved from Manchester to become a chemistry professor at Oxford University in January 1913. He supervised a variety of wartime organic chemistry projects in the university's chemistry laboratory for government bodies such as the Chemical Warfare Committee and the Department of Explosive Supplies.[27] One of projects focused on improving the process for converting phenol into picric acid, an organic chemical, also known as trinitrophenol, that was used to make Lyddite, a high explosive used by the British Army early on in the war.

The largest team carrying out war work in Perkin's department investigated the synthesis of acetone from ethanol, an alcohol that could be readily produced from sugars by fermentation followed by distillation. The proposed synthesis involved three stages each of which used a catalyst to speed up the reaction: first conversion of ethanol to acetaldehyde (also known as ethanal) by partial oxidation; then oxidation of the aldehyde to acetic acid (ethanoic acid); and finally the production of acetone from the acid.

The synthesis had been developed in the laboratory to a large extent by Bertram Lambert (1881–1963) who, like Sidgwick, was a demonstrator in chemistry at the university. After the war broke out, Lambert left Oxford to join the army as an officer in the Royal Engineers. Sidgwick then took over responsibility for the project.

A plant to produce acetone on an industrial scale by the three-stage process was built at the Joseph Crosfield and Sons soap and chemical manufacturing works at Warrington, a town in north-west England.[28] However, scaling up the third stage of the process proved problematic and it was not possible to obtain satisfactory yields of acetone from the acetic acid. Furthermore, an explosion at the plant while the work was in progress killed several men, Tizard noted.[26] Even so, "the process was used throughout the war for the manufacture of acetic acid," he added.

Acetic acid had numerous uses in the war including the manufacture of the acetic anhydride that was used to make cellulose acetate photographic films and airplane dopes. Acetic anhydride was also employed to convert morphine into diacetylmorphine, more commonly known as diamorphine or heroin. Both morphine and diamorphine were widely used as painkillers in the First World War.

OTHER ROUTES

On the other side of the Atlantic, the American inventor and organic chemist Carleton van Staal Ellis (1876–1941) was busy developing processes for converting the more volatile hydrocarbon components of petroleum into ketones, notably acetone, and alcohols. The distillation of crude oil produces a series of fractions each of which boils over a specific temperature range. One of the fractions, generally known as naphtha, contains hydrocarbons with up to 12 carbon atoms in each molecule.

Ellis not only developed a method known as the "tube-and-tank" process for cracking, that is breaking up, the hydrocarbon molecules in the naphtha into molecules with smaller numbers of carbon atoms, he also discovered a way of converting the alkenes obtained by the cracking process into alcohols. Alkenes, also known as olefins or olefines, are hydrocarbon compounds consisting of molecules with open chains of carbon atoms with at least one carbon–carbon double bond. In 1916, Ellis showed that alcohols could be manufactured by first dissolving the alkenes in sulfuric acid and then hydrolysing the products by heating them with water or treating them with steam. The alcohols were finally oxidised, typically by using chromic acid, to form the corresponding ketones.

One of the simplest alkenes produced by Ellis's cracking process was propene (or propylene). Each molecule of the alkene has a chain of three carbon atoms connected by one double bond and one single bond. The propene was then converted into isopropyl alcohol and then into the acetone which the United States used to make airplane dope following its entry into the war.

Figure 7.6   Carleton Ellis (William Haynes Portrait Collection, Chemical Heritage Foundation Image Archives, Philadelphia, PA).

In Canada, acetone was manufactured from another source of organic chemicals, namely coal or more specifically, the product of its destructive distillation: coke. The coke was first converted into calcium carbide which in turn was converted into acetylene (see Chapter 5). Oxidation of the acetylene yielded acetaldehyde and then acetic acid. The acid was then passed over a hot catalyst to produce acetone,[29] using a similar process to the one that Sidgwick and Joseph Crosfield & Sons had attempted but failed to develop.

The Germans employed a similar route to make acetone from coke, except for the final stage. Rather than heating the acetic acid over a catalyst, they converted the acid into calcium acetate in the same way that wood vinegar was converted into the acetate (see above). They then dry distilled the acetate to produce acetone. Much of the acetone was employed to manufacture a synthetic rubber known as methyl rubber.[30] Germany had turned to the production of this type of artificial rubber, even though it was inferior in quality to natural rubber, when the British Royal Navy blockade halted German imports of natural rubber and other essential raw materials. The country needed the rubber to make tyres, hoses, and other products essential for its war effort.

RINTOUL'S RECOVERY PROCESS

When the war broke out, Weizmann wrote to the British War office suggesting that it might be interested in developing his fermentation process to make acetone. He did not receive a reply. He then contacted the Nobel Explosives Company in Ardeer, Scotland, and in February 1915 received a visit from the company's chief chemist, William Rintoul (1870–1936).

Rintoul, a Scottish chemist, was an explosives expert. In 1894, he had taken up an appointment at the Royal Gunpowder Factory at Waltham Abbey, a market town some 15 miles north of London. Six years later, he was put in charge of the manufacture of nitroglycerine and soon became chief chemist at the factory. He moved to Ardeer in 1909.[31]

In his autobiography, Weizmann claimed that he initially did not know the purpose of Rintoul's visit.[32] The two chemists gossiped about the war and then Rintoul expressed interest in Weizmann's research. He was impressed by Weizmann's fermentation process and recommended it to Frederick Nathan (1861–1933) who was then advising the Admiralty on cordite supply. Nathan, a colonel in the British Army's Royal Artillery, joined the Waltham Abbey factory in 1889 and was subsequently given responsibility for the development of cordite for the British armed forces. He was engaged as manager of the Ardeer explosives works in 1909 and after war broke out became advisor to the Admiralty on cordite supply. In 1916, he was appointed Controller of Propellant Supplies under the Ministry of Munitions and consequently assumed responsibility for the manufacture of acetone in Britain.

Rintoul's visit to Weizmann and his contact with Nathan sparked the chain of events that eventually led to the mass production of acetone by the fermentation process in Britain, Canada and the United States. Rintoul, however, also made another significant contribution to the availability of acetone for cordite production that was totally unrelated to the Weizmann process.

Cordite was prepared as a gelatinous mixture of nitrocellulose, nitroglycerine and mineral jelly using acetone as a solvent. The mixture was then kneaded into cords, but as the cords retained much of the solvent they were loaded into driers to drive off the solvent. When the propellant was first manufactured, about two tons of acetone was lost through evaporation into air for every ten tons of cordite produced.

Rintoul found a way of recovering the acetone vapours from air. The process essentially involved passing the air–acetone mixture into an aqueous solution of sodium bisulfite (also known as sodium hydrogen sulfite) in a scrubbing tower. The solution absorbed the acetone which then combined with the sodium compound to form acetone sodium bisulfite. Distillation of the solution yielded acetone as a distillate that could be recycled in the manufacture of cordite.

Rintoul worked on the project at Waltham Abbey with another Scottish chemist and explosives expert, Robert Robertson (1869–1949). The two chemists, who became close friends, described the process in a patent published in 1903.[33] Some three decades later, Rintoul married Jess Isabel, one of Robertson's three younger sisters. She was his second wife.

ROBERTSON'S SOLUTION TO ACETONE SCARCITY

The scarcity of acetone presented a major challenge to cordite manufacture in Britain in the early years of the war. Weizmann's fermentation process subsequently helped to meet the increasing demands for the solvent, but the question soon arose: why use acetone at all? Was it not possible to find a different and more widely available solvent that could be used to make a propellant with the same ballistic performance as cordite?

Robertson and his wartime team of research chemists at the Royal Gunpowder Factory at Waltham Abbey had the answer. The Scottish chemist had considerable experience of working not only on propellants but also on the high explosives that were used to fill artillery shells. In 1892, he had obtained a post as a junior chemist at the factory and spent a couple of years in the laboratory carrying out experiments on nitrocellulose, one of the

Figure 7.7 Robert Robertson (Royal Society of Chemistry Library).

components of cordite. Many of his experiments focused on the solubility of various types of nitrocellulose in mixtures of diethyl ether and ethanol, the type of solvent mixture that the French used to make their *Poudre B* propellant.

After his spell in the laboratory, Robertson supervised the manufacture of nitroglycerine, another component of cordite, at the Waltham Abbey factory. At the same time, he also worked with Rintoul on the acetone recovery process. When he was put in charge of the main laboratory at the factory in 1900 he began studying the stability and purification of nitrocellulose and nitroglycerine.

In 1907, having spent several months in India with Nathan examining the safety and stability of cordite in hot climates, he was appointed head chemist in the chemical research department of the Royal Arsenal at Woolwich in south-east London. Over the following years, the department carried out research on the synthesis and properties of TNT and other explosives.

The team synthesised and subsequently perfected a process for the manufacture of the high explosive tetryl.

When the war started, nine chemists and support staff worked in the department's laboratories under the supervision of Robertson who by now had been promoted to the position of Director of Explosives Research. One of the team's most urgent problems was to find ways of improving the efficiency of the processes used to make TNT in sufficient quantities to meet the increasing demands of the British armed forces. At the time, these processes were slow and inefficient. The work at Woolwich soon led to a significant increase in TNT production but there was a problem. The toluene needed to make the high explosive was in short supply. Robertson and his staff found a solution. They showed that supplies of TNT could be eked out by "diluting" it with another explosive, ammonium nitrate. The mixture, named "amatol," had similar explosive properties to pure TNT, and was subsequently used extensively as a bursting charge in British high explosive shells. Furthermore, the mixture was safer to handle than TNT and less expensive to produce because ammonium nitrate was easily manufactured at low cost in large quantities.

Amatol was "the greatest single thing of importance" in boosting the supply of munitions in Britain and enabled the British Army "to expend shell on an unlimited scale," observed Lord John Fletcher Moulton (1844–1921) after the war.[34] Moulton, a lawyer, was in charge of the development and manufacture of high explosives in the country during the war.

The supply of propellants was just as important, however. The army could not fire shells filled with TNT or amatol without propellants. Robertson's group knew this and so, well before the development of the Weizmann process, the chemists began to tackle the "why use acetone to make cordite?" question. The standard cordite used at the time contained guncotton which, as we saw above, is a highly nitrated form of cellulose. Guncotton, although soluble in acetone, was insoluble in the mixture of diethyl ether and ethanol that Robertson had examined many years earlier. Collodion was soluble in the mixture, however.

After exhaustive studies of the use of collodion as a substitute for guncotton and the diethyl ether–ethanol solvent mixture as a replacement for acetone, Robertson and colleagues devised a formula for a new propellant with similar properties to the acetone-based cordite. It was named Cordite RDB, the abbreviation standing for Research Department formula B. The propellant consisted of 52% collodion, 42% nitroglycerine and 6% petroleum jelly.

The Royal Gunpowder Factory and His Majesty's Factory at Gretna, a town in Scotland that lies on the border with England, started producing vast quantities of the new propellant in late 1916. British distilleries supplied the alcohol, amounting to hundreds of tons each week, required for its manufacture. The Gretna factory was one of the largest explosive factories in the British Empire. Construction of the factory began in November 1915 and production of the propellant started in April the following year. It employed well over 11 000 women and 5000 men and had the capacity to produce around 800 tons of Cordite RDB each week.

The new propellant was used mainly by the British Army. However, it had the drawback that it tended to become unstable when stored for long periods. The Royal Navy preferred the standard acetone-based cordite as it was more stable and more uniform in its effect. With the increased availability of acetone towards the end of the war, the navy reverted to using the older form of cordite.

By the time the war ended, the number of scientists in Robertson's department had risen to 110.[35] Robertson was elected a Fellow of the Royal Society in 1917 and honoured with a knighthood the following year in recognition of his services during the war.

REFERENCES

1. G. I. Brown, *Explosives: History with a Bang*, The History Press, Stroud, 2010, p. 168.
2. War Office, *Treatise on Ammunition*, 10th edn, The Naval & Military Press, Uckfield, East Sussex, and The Imperial War Museum, London, 1915, p. 12.
3. R. MacLeod, The chemists go to war: the mobilization of civilian chemists and the British war effort, 1914–1918, *Ann. Sci.*, 1993, **50**, 455.

4. M. Gorman and C. Doering, History of the structure of acetone, *Chymia*, 1959, **5**, 202.
5. War Office, *Treatise on Ammunition*, 10th edn, The Naval & Military Press, Uckfield, East Sussex, and The Imperial War Museum, London, 1915, p. 26.
6. S. F. Acree, What chemistry has done to aid the utilization of wood, *Ind. Eng. Chem.*, 1915, 7, 913.
7. R. B. Pilcher, *What Industry Owes to Chemical Science*, The Scientific Book Club, London, 1947, p. 228.
8. J. R. Withrow, The chemical engineering of the hardwood distillation industry, *Ind. Eng. Chem.*, 1915, 7, 912.
9. Hansard, 15 June 1910, Commons Sitting - Oral Answers to Questions - Acetone (Government Supply). See: http://hansard.millbanksystems.com/commons/1910/jun/15/acetone-government-supply.
10. Hansard, 28 March 1911, Commons Sitting - Oral Answers to Questions - Acetate of Lime and Acetone. See: http://hansard.millbanksystems.com/commons/1911/mar/28/acetate-of-lime-and-aceteone.
11. History Extra, Did conkers help win the First World War? 22 October 2014, http://www.historyextra.com/conker.
12. C. Weizmann, *Trial and Error: The Autobiography of Chaim Weizmann*, Hamish Hamilton, London, 1949, p. 16.
13. C. Weizmann, *Trial and Error: The Autobiography of Chaim Weizmann*, Hamish Hamilton, London, 1949, p. 99.
14. C. Weizmann, *Trial and Error: The Autobiography of Chaim Weizmann*, Hamish Hamilton, London, 1949, p. 127.
15. B. Litvinoff, *The Essential Chaim Weizmann: The Man, the Statesman, the Scientist*, Weidenfeld and Nicolson, London, 1982, p. 3.
16. D. T. Jones and D. R. Woods, Acetone-butanol fermentation revisited, *Microbiol. Rev.*, 1986, **50**, 484.
17. C. Weizmann, *Trial and Error: The Autobiography of Chaim Weizmann*, Hamish Hamilton, London, 1949, p. 171.
18. C. Weizmann, *Trial and Error: The Autobiography of Chaim Weizmann*, Hamish Hamilton, London, 1949, p. 220.
19. Anon., Obituary, *J. Proc. R. Inst. Chem.*, 1949, 73, 310.
20. C. Weizmann, Production of acetone and alcohol by bacteriological processes, *US Pat.* 1 315 585, 1919.

21. B. Litvinoff, *The Essential Chaim Weizmann: The Man, the Statesman, the Scientist*, Weidenfeld and Nicolson, London, 1982, p. 10.
22. E. D. Bergmann, Obituary notice, Chaim Weizmann, 1874–1952, *J. Chem. Soc.*, 1953, 2840.
23. C. Weizmann, *Trial and Error: The Autobiography of Chaim Weizmann*, Hamish Hamilton, London, 1949, p. 440.
24. N. V. Sidgwick, *Organic Chemistry of Nitrogen*, Clarendon Press, Oxford, 1910.
25. L. E. Sutton, Obituary notices: Nevil Vincent Sidgwick, 1873–1952, *Proc. Chem. Soc.*, 1958, 297.
26. H. T. Tizard, Nevil Vincent Sidgwick, 1873–1952, *Obit. Not. Fell. R. Soc.*, 1954, **9**, 236.
27. J. Morrell, Research as the thing: Oxford chemistry 1912–1939, in *Chemistry at Oxford: A History from 1600 to 2005*, ed. R. J. P. Williams, J. S. Rowlinson, and A. Chapman, Royal Society of Chemistry, Cambridge, 2009, p. 137.
28. G. Hartcup, *The War of Invention: Scientific Developments, 1914–18*, Brassey's Defence Publishers, London, 1988, p. 51.
29. American Chemical Society, *National Historic Chemical Landmarks, Discovery of the Commercial Processes for Making Calcium Carbide and Acetylene*, American Chemical Society, Washington, DC, 1998.
30. R. W. Thomas and F. Farago, *Industrial Chemistry*, Heinemann, London, 1973, p. 188.
31. Anon., Obituary: William Rintoul, *J. Proc. Inst. Chem. GB Irel.*, 1936, **60**, 333.
32. C. Weizmann, *Trial and Error: The Autobiography of Chaim Weizmann*, Hamish Hamilton, London, 1949, p. 218.
33. R. Robertson and W. Rintoul, Manufacture of explosives, celluloid, or the like, *US Pat.* 723 311 A, 1903.
34. G. Hartcup, *The War of Invention: Scientific Developments, 1914–18*, Brassey's Defence Publishers, London, 1988, p. 49.
35. R. C. Farmer, Robert Robertson: 1869–1949, *Obit. Not. Fell. R. Soc.*, 1949, **6**, 539.

CHAPTER 8

Whaling for War

AN ESSENTIAL CONTRIBUTION

The discovery that whaling made an essential contribution to the First World War came as a surprise to me. I knew that troops rubbed whale oil into their feet to prevent trench foot. Indeed, I had mentioned that in my previous book on the chemistry of the war, published in October 2012.[1]

Following publication of the book, I began to look further into the sources of chemicals used in the war. It was only then that I became aware that whale oil was a prime source of the glycerine used to make nitroglycerine, a component of the propellant cordite and the blasting explosive dynamite.

The industrial-scale killing and wounding of soldiers and civilians in the war required the industrial-scale production of such explosives. So it soon became obvious to me that this production depended, to a large extent, on the industrial-scale slaughter of whales.

A SHORT HISTORY OF WHALING

Hunting and killing whales dates back at least two thousand years when hunters in boats herded whales swimming in coastal waters into shore before killing them. By the 12th and 13th

The Chemists' War, 1914–1918
By Michael Freemantle
© Freemantle, 2015
Published by the Royal Society of Chemistry, www.rsc.org

centuries, Basque whalers were operating on an industrial scale along the coast of the Bay of Biscay. They initially hunted North Atlantic right whales, a species of baleen whales that feed on small marine organisms, primarily plankton such as krill. Right whales are fairly slow swimmers and unlike other species of baleen whales, they float after being killed making it easy for sailors to haul the dead whales alongside their whaling vessels.

Baleen is a type of keratin—a fibrous protein that occurs throughout the animal kingdom. The protein is found in various forms in our hair, skin, and nails, and also in wool, silk, hooves, claws, horns, beaks and other parts of animals. Baleen whales employ plates of bristle-like baleen hairs in their mouths as filters to strain the marine organisms from water.

During the 14th century increasing demand for whale products and declining stocks in the Bay of Biscay encouraged the Basque whalers to hunt as far north as Iceland and the Gulf of Saint Lawrence. By the 17th century, whalers from several European countries were hunting Arctic right whales and other species of baleen whales, such as the bowhead whale, along the ice edge of the North Atlantic and Arctic.

In the 18th and 19th centuries American whalers gradually began to play an important role in the whaling industry. The vessels hunted sperm whales and other species of whales in the South Pacific and as far away as Australia. Sperm whales, unlike baleen whales, are predators. They have jaws and sharp teeth that enable them to feed on fish and squid.

At the outbreak of the First World War almost 70% of the whaling in the world was carried out in the seas around groups of islands in the South Atlantic and Antarctic known as the Falkland Island Dependencies.

By 1914, a revolution had taken place in whaling methods. Until the middle of the 19th century, whaling was typically carried out using hand harpoons. Sailors in wooden boats impaled whales at close range with hand harpoons attached to ropes. The sailors then killed the whales with lances. Japanese whalers also caught whales in nets.

In the 1860s, Norwegian Svend Foyn (1809–1894), a son of a ship-owner, developed and introduced the grenade harpoon gun that could be fired from a steam-powered boat. The harpoon was

BOATS ATTACKING WHALES.

Figure 8.1   "Boats attacking whales" is from Thomas Beale's *The Natural History of the Sperm Whale*, London, 1839 (Philip Hoare Collection).

fired at the whale from a gun positioned on the prow of a boat. The invention allowed sailors to systematically hunt and kill rorquals, a species of baleen whale that is the largest and fastest swimming of all whales. The harpoon hooked into the whale and within seconds the grenade exploded, more often than not resulting in a slow, painful, and bloody death. The invention heralded the beginning of modern whaling.

Virtually every part of the whale was used for one purpose or another. Ambergris, a solid waxy material secreted in the intestines of sperm whales, was highly prized as a stabilizer in the manufacture of perfumes. Spermaceti, a white waxy oil extracted from the heads of sperm whales, was used to make candles, cosmetics, ointments, and lubricants. Whale meat was consumed throughout the ages, cartilage was converted into glue, and whale skin was used to make leather. Baleen, a strong and flexible material, was employed as stays in women's corsets and to make the ribs of umbrellas. Before the advent of synthetic plastics in the late 19th and early 20th centuries, baleen was the most commercially important part of the whale.

WHALE OIL FROM BLUBBER

Whale oil prepared from blubber, the layer of fatty tissue beneath the whale's skin, became increasingly valued during the 19th and early 20th centuries. It was used in oil lamps, as a lubricant, for tanning, and as a raw material for the manufacture of varnishes, paints, linoleum, soaps, margarine, and explosives.

The layer of blubber, which could be as much as 50 cm thick, was cut and removed from the whale in a process known as flensing. It was then rendered into oil by boiling and melting strips of the blubber in open pans. Each whale typically produced three or four tons of oil, although the yield varied immensely depending on a number of factors, not least the species and size of the whale.

In many cases there was immense waste, especially when the whale harvest was abundant. It was not unusual for an on-shore whaling station or floating factory ship to strip only the more accessible dorsal blubber from the whale and not the abdominal blubber. During whaling operations in the Antarctic from 1908

Figure 8.2   Flensing. The plate is from a Victorian edition of a whaling voyage: William M. Davis, *Nimrod of the Sea; or, The American Whaleman.* Harper & brothers, New York, 1874 (Philip Hoare Collection).

to 1914, whaling companies sometimes "literally wallowed in whales, selecting only the best blubber and allowing the carcass, with 60% of its content, to drift away," note Norwegian historian Arne Odd Johnsen (1909–1985) and co-author J. N. Tønnessen in their book *The History of Modern Whaling*.[2] They add that at the beginning of the 1912 whaling season, some 6000 whale carcasses were lying around on the beach at Deception Island in Antarctica.

The blubber of baleen whales contains between 50 and 80% oil by weight. Right whales are particularly fat and therefore produced high yields of oil. Furthermore, they have long baleen plates hanging from their upper jaws. These factors combined with their propensity to swim slowly and close to shore and float when killed made them the "right" whales to catch; hence their name.

Whale oil is essentially a mixture of triglycerides, a family of organic chemicals that occur not only in blubber but also in other animal fats and vegetable oils. Each triglyceride molecule consists of three fatty acid components attached to the spine of a glycerol molecule. Fatty acids are organic acids with long chains of carbon atoms. Glycerol itself is a simple alcohol with three hydroxyl groups.

Soap, one of the most ancient of all chemically-manufactured materials, is traditionally made from these naturally-occurring triglyceride-containing fats and oils in a process known as saponification. The process typically involves boiling the fat or oil with an aqueous solution of an alkali such as sodium hydroxide or potassium hydroxide. The product, soap, is the sodium or potassium salt of the fatty acid. It is separated from the aqueous liquors which contain glycerol as a by-product.

Until the beginning of the 20th century, whale oil was inedible. Then, in 1907, a small quantity of whale oil was hydrogenated for the first time to produce a hardened fat that could be used to make margarine. The process converted carbon–carbon double bonds in the fatty acid chains to carbon–carbon single bonds. The industrial production of edible fat from whale oil began in 1910.

This development together with other factors, particularly the price and availability of cattle fats and vegetable oils, eventually led to a "fat war" with soap and margarine manufacturers competing for the purchase of fats and oils, including whale oil. In 1913, leading European companies, such as the British soap maker Lever Brothers, established the Whale Oil Pool for the fair distribution of the oil from producers to consumers.

From 1914 onwards, whale oil was a key strategic material for the conduct of the Great War. It was used as rifle oil, as a fuel for trench stoves, to protect against trench foot, and above all for the production of the vast quantities of the glycerol, or glycerine as it is also known, needed for the manufacture of explosives.

TRENCH FOOT

Along the Western Front during the cold wet winter of 1914/ 1915, many soldiers wearing non-waterproof boots had to walk along muddy roads and stand or sit in near-freezing waterlogged and poorly drained trenches, sometimes for days if not weeks on end. Their boots soon became sodden and their feet cold and damp.

This prolonged exposure to mud and water underfoot led to a medical condition known as "trench foot." Symptoms included blue feet, lesions, swelling, numbness, blackening of the toes, and skin on the feet dying and peeling off. Furthermore, the feet often became infected with pathogenic microorganisms that thrived in the mud leading to diseases such as tetanus.

"Your feet swell up and go completely dead. You could stick a bayonet into them and not feel a thing. If you are fortunate enough not to lose your feet and the swelling begins to go down, it is then that the agony begins. I have heard men cry and even scream with the pain and many had to have their feet and legs amputated." This vivid description of trench foot by Harry Roberts, a sergeant in the Royal Army Medical Corps (RAMC), is one of many first-hand accounts of life, death and suffering of British soldiers in the trenches collected together and published in *Tommy Goes to War*.[3] The accounts were extracted from their diaries and letters.

Trench foot was treated with antiseptics and analgesics such as aspirin. Injections of anti-tetanus serum were also given to soldiers suffering from the condition. In some cases where trench foot was left untreated, gangrene set in and surgeons had to amputate one or more toes. During the war, more than 70 000

British soldiers serving on the Western Front suffered from the condition, 75% of whom were sent back to base for treatment, according to historian Denis Winter.[4]

As the war progressed, medical officers soon realised that preventative as well as curative measures were necessary to tackle the condition. They included the provision of knee-length rubber boots and regular foot inspections by officers. Soldiers were ordered to dry their feet at least twice a day, rub in whale oil, and put on dry socks. They did not enjoy the task, however, as the whale oil had a foul fishy odour, but they had no choice. "The army made the disease a crime," Winter writes.

GLYCERINE

Carl Wilhelm Scheele (1742–1786), a self-taught Swedish chemist, discovered glycerine in 1779 when he heated litharge, a lead oxide mineral, with olive oil. The sweet-tasting syrupy liquid he extracted from the mixture was subsequently named "glycerine" after the Greek word for "sweet." A few years later, Scheele showed that the oily liquid could be produced not only from other vegetable oils but also from animal fats.

Although it had been produced as a by-product in the manufacture of soap from oils and fats for centuries, glycerine had few uses until the discovery of nitroglycerine in 1847. By the time of the First World War, the chemical had become the main product of soap factories rather than a by-product.

However, the demands for the material to make explosives during the war far exceeded supplies from the soap industries. Additional supplies were also produced commercially from sugar using a yeast fermentation process.[5]

Nowadays, glycerine is produced from vegetable oils such as coconut oil and palm oil, from fish oils, and from tallow, which is the solid rendered fat of cattle and sheep. The compound is also produced chemically from propene, an organic chemical extracted from fossil fuels, and by other chemical routes. It is used as a starting material to synthesize numerous chemicals and also employed extensively not only to manufacture explosives but also in foodstuffs, personal care products, pharmaceutical preparations, antifreeze mixtures and a whole variety of other applications.

Figure 8.3   Carl Wilhelm Scheele (Royal Society of Chemistry Library).

NITROGLYCERINE

Spelt nitroglycerin in the United States, nitroglycerine is also known as glyceryl trinitrate, nitroglycerol, and by a variety of other names. The pale yellow oily liquid was a component of a variety of explosives employed during the First World War. They included the British propellants cordite and ballistite, the German propellant *Rohr-Pulver* (tube powder), and the blasting explosives dynamite and gelignite.

Nitroglycerine is a high explosive which means that, unlike low explosives such as gunpowder, the shock wave of the explosion travels faster than the speed of sound. As with other high explosives, for example nitrocellulose (otherwise known as guncotton) and TNT, nitroglycerine derives its explosive power from the high proportion of nitrogen and oxygen in the compound. When pure, nitroglycerine is unstable and detonates if rapidly heated or given a sharp blow. Detonation results in the production of gases with pressures up to 50 times greater than those produced when gunpowder is ignited. The blast of a nitroglycerine explosion is so powerful it would destroy a gun barrel if used as a propellant by itself. The explosive is therefore incorporated into other materials to ensure it is safe to handle, transport, and use.

The compound is made by adding pure glycerine slowly to a mixture of concentrated nitric and sulfuric acids at a temperature just above the freezing point of water. It was discovered in 1846 by Ascanio Sobrero (1818–1888), an Italian organic chemist who had been a student of French chemist Théophile-Jules Pelouse (1807–1867) in Paris. Pelouse had carried out experiments on nitrocellulose and related chemicals in his laboratory. It is notable that another of his students was Alfred Nobel (1833–1896), the Swedish chemist and industrialist who was to make a fortune from his patented inventions of dynamite and other explosives.

Sobrero was appointed professor of chemistry at the University of Turin, Italy, in 1845. While carrying out research on explosives, he heated a drop of nitroglycerine in a test tube. It exploded, the splinters of glass wounding and scarring his hands and face.[6] He subsequently considered nitroglycerine to be too dangerous for any practical use and did not report the discovery until 1847.[7] "When I think of all the victims killed during nitroglycerine explosions, and the terrible havoc that has been wreaked, which in all probability will continue to occur in the future, I am almost ashamed to admit to be its discoverer," he said.[8]

Although Nobel always openly recognized Sobrero's discovery, the Italian chemist is said to have been mortified when Nobel began to develop and exploit the explosive commercially. On 3 September 1864, Nobel's youngest brother, 20-year-old Emil, and five workers were killed in a nitroglycerine explosion that destroyed the Nobel factory in Stockholm. The disaster so traumatized their father, Immanuel Nobel (1801–1872), that he had a stroke and remained bedridden for the rest of his life.[9] The tragedy prompted Alfred Nobel to look for ways of making the high explosive safe to handle. In 1866, he showed that it could be stabilized by mixing it with diatomaceous earth, an inert naturally-occurring porous solid, otherwise known as diatomite or kieselguhr. He patented the mixture under the name "dynamite" in 1867.

Apart from being a widely-used explosive, nitroglycerine is also valuable as a drug for treating heart conditions. Indeed it has been called the "explosive drug."[6] The first person to use it for medicinal purposes was London-born doctor William Murrell (1853–1912). In the summer of 1878, he began to administer nitroglycerine to patients suffering from angina, reporting its use and efficiency in several papers published in the journal *Lancet* the following year.[10] The drug is a vasodilator which means that it increases the diameter of arteries and other blood vessels. During the First World War, it was used by RAMC (Royal Army Medical Corps) medical officers to treat high blood pressure in troops suffering from diseases such as trench nephritis, that is inflammation of the kidneys induced by living in the trenches.[11]

IMMENSE DEMAND FOR WHALE OIL

In poor weather during the First World War when trenches were drenched with mud and water, a battalion of up to 1000 soldiers would use between 40 and 50 litres of whale oil each day as protection against trench foot. The demand for the oil for this purpose, however, paled into insignificance when compared with the demands to make propellants and other explosives for the armed forces. Every single round of firearm ammunition, whether for a pistol, a rifle, or a machine gun, required a propellant, and every single artillery shell was fired by a charge of propellant contained in a bag or a brass cartridge case.

The standard British propellant used during the war was cordite. It was known as a smokeless propellant because, unlike gunpowder, it produced relatively little smoke when fired. Early types of cordite consisted of 58% nitroglycerine, 37% guncotton (nitrocellulose), and 5% petroleum jelly. In a later version, known as Cordite MD, where MD is an abbreviation of MoDified, the proportions of nitroglycerine and guncotton were changed to 30% and 65%, respectively. Cordite RDB, named after Research Department formula B developed at the Royal Arsenal at Woolwich in south-east London, contained 42% nitroglycerine,

52% of a form of nitrocellulose known as collodion, and 6% petroleum jelly (see Chapter 7).

The amount of cordite needed to fire a single shell varied enormously depending on the type of gun and type of shell. For example, the Mark 2 version of the 9.2-inch breech-loading siege howitzer used by British forces in France during World War I required a charge of 10.76 kg of Cordite MD to fire a single high explosive shell weighing 131 kg.[12] The Royal Field Artillery's standard field gun, the quick-fire (QF) 18-pounder, on the other hand, required a cartridge filled with just 650 g of Cordite MD to fire an 18.5 pound (8.39 kg) shrapnel shell. A simple but approximate calculation shows that at least 1 kg of nitroglycerine was needed to produce enough Cordite MD to fire five 18.5 pound shrapnel shells. Firing a million 18.5 pound shells therefore required propellant containing around 200 000 kg of nitroglycerine.

During the Battle of the Somme, fought between July and November 1916, the British and German armies fired some 30 million shells of various types at one another, averaging about 150 per minute. The British QF 18-pounder gun alone fired 86 million 18.5 pound shrapnel, gas, high explosive and other types of shells during the war. Altogether, the British Army and Royal Navy fired almost 250 million shells of various types and weight, which is roughly an average of 100 shells every minute throughout the war.

The artillery warfare in France and Belgium consequently placed immense demands on the munitions industries. In 1917, for example, production of Cordite RDB at His Majesty's Factory in Gretna, a town that lies in Scotland on the border with England, reached a peak production of 800 tons per week. That required well over 300 tons of nitroglycerine weekly which in turn required over 120 tons of glycerine.

The demand for vegetable oils, animal fats, and not least whale oil in the war to produce the glycerine needed to make nitroglycerine was therefore enormous. Mineral oil could not be used at the time. The oil is extracted from non-vegetable or non-animal sources such as petroleum and consists of hydrocarbons rather than triglycerides. It was not until after the war that industrial processes were developed to convert propene, a fossil fuel hydrocarbon, into glycerine.

Figure 8.4 British 18 pounder field gun (author's own, 2011, Imperial War Museums, London).

SUPPLIES OF WHALE OIL DURING THE WAR

According to Tønnessen and Johnsen, whale oil became particularly important during the Great War not just for the manufacture of nitroglycerine, but also because "supplies of vegetable oils and fats failed."[13] Before the war, whaling was dominated by Britain and particularly Norway. The two countries accounted for approximately 84% of the total catch in the 1912–1913 whaling season, for example. Almost 70% of the whaling operations were carried out in British waters around the Falkland Island Dependencies. Norway sold over 30% of its whale oil to Germany and Austria, which were chronically short of animal fats, and also a substantial amount to Britain.[14]

Being a neutral country, Norway was entitled to sell its whale oil to both the Central Powers and the Allied Powers during the war. However, the country relied on Britain for coal. After war broke out, Britain threatened to cut off supplies of fuel to Norway and refuse coal-bunkering facilities for its whaling ships if it continued to supply Germany and Austria with whale oil. In 1916, Britain declared whale oil contraband, banned whale exports from the Falkland Island Dependencies to any country other than the United Kingdom, and cancelled Norwegian whaling licences.

The restrictions were relaxed later in the year and Norway was "persuaded" to sell its whale oil to British companies at the knock-down price of £50 per ton whereas Germany was offering £300 per ton. "There was more than a grain of truth in the saying that Norway was forced to produce cheap explosives for Britain," Tønnessen and Johnsen note.[15]

Apart from supplies of oil from its own whaling operations and those from Norway, Britain also had an additional supply from an unlikely source: the Dutch margarine and soap manufacturer Anton Jurgens (1867–1945), whose grandfather had built the world's first margarine factory in 1871. Jurgens anticipated that there would be a shortage of oils and fats in Europe following the war and that a substantial profit could be made by stockpiling whale oil for his margarine and soap factories. In 1915, he began to buy large quantities of the oil, storing most of it in Britain. The British Board of Trade subsequently offered to buy the oil to produce glycerine but Jurgens refused the offer. In response, the British government requisitioned the oil and paid for it at the offer price.

Germany did not experience any significant shortfalls in the supplies of oil for the production of glycerine until the autumn of 1916.[16] It then rapidly introduced a sugar fermentation process and within a few months was producing the amounts of glycerine it needed for manufacturing propellants.

The United States also played a key role in the supply of oils, fats, and explosives to Europe during the war. "The Allies would have been much harder pressed much earlier in the war had it not been for the American chemist," observes O. P. Hopkins, a reporter in Washington, DC, in an article published in 1918.[17] He records that exports of dynamite, for example, rose from 14.46 million pounds (6.56 million kg) in 1914 to almost 18.91 million pounds (8.58 million kg) in 1918. The export of stearin, a triglyceride found in many animal and vegetable fats, increased from 2.72 million pounds (1.23 million kg) to 10.25 million pounds (4.65 million kg) over the same period.

Even though sugar fermentation and a variety of animal and vegetable oils and fats were exploited for the manufacture of glycerine during the war, whale oil was the prime source for the production of nitroglycerine explosives. According to one source, the total world catch from just before the war until the end of the war totalled almost 80 000 whales.[18]

If it had not been for the war, oil from these whales would have almost certainly been used for soap and margarine manufacture and for other purposes. Even so, there is no question that tens of thousands of whales were slaughtered during the war for the production of the propellants that led to the slaughter of millions of soldiers.

REFERENCES

1. M. Freemantle, *Gas! GAS! Quick, boys! How Chemistry Changed the First World War*, The History Press, Stroud, 2012, p. 184.
2. J. N. Tønnessen and A. O. Johnsen, *The History of Modern Whaling*, C. Hurst & Company, London, and Australian National University Press, Canberra, 1982, p. 184.
3. M. Brown, *Tommy Goes to War*, The History Press, Stroud, 2009, illustration 31 caption.
4. D. Winter, *Death's Men*, Penguin Books, London, 1979, p. 99.
5. Z.-X. Wang, J. Zhuge, H. Fang and B. A. Prior, Glycerol Production by Microbial Fermentation: A Review, *Biotechnol. Adv.*, 2001, **19**, 201.
6. L. C. Holmes and F. J. DiCarlo, Nitroglycerin: the explosive drug, *J. Chem. Educ.*, 1971, **48**, 573.
7. A. Sobrero, Sur plusieurs composes détonants produits avec l'acide nitrique et le sucre, la dextrin, la lactine, la mannite et las glycerine, *De Compt. Rendus*, 1847, **24**, 247.
8. J. Wisniak, Ascanio Sobrero, *Revista CENIC. Ciencias Quimicas*, 2006, **37**, 41.
9. G. I. Brown, *Explosives: History with a Bang*, The History Press, Stroud, 2010, p. 112.
10. W. Murrell, Nitro-glycerine as a remedy for angina pectoris, *Lancet*, 1879, **113**, 80, 113, 151, 225.
11. J. M. Clarke, Trench nephritis, its later stages and treatment, *Br. Med. J.*, 1917, **2**, 239.

12. I. V. Hogg and L. F. Thurston, *British Artillery Weapons & Ammunition 1914–1918*, Ian Allan, Shepperton, 1972, p. 245.
13. J. N. Tønnessen and A. O. Johnsen, *The History of Modern Whaling*, C. Hurst & Company, London, and Australian National University Press, Canberra, p. 295.
14. S. J. Holt, Sharing the Catches of Whales in the Southern Hemisphere, in Case Studies on the Allocation of Transferable Quota Rights in Fisheries, Ed., R. Shotton, *FAO Fisheries Technical Paper*, No. 411, FAO, Rome, 2001, p. 322.
15. J. N. Tønnessen and A. O. Johnsen, *The History of Modern Whaling*, C. Hurst & Company, London, and Australian National University Press, Canberra, p. 299.
16. J. A. Johnson, Technological mobilization and munitions production: comparative perspectives on Germany and Austria, in *Frontline and Factory: Comparative Perspectives on the Chemical Industry at War, 1914–1924*, ed. R. MacLeod and J. A. Johnson, Springer, Dordrecht, 2006, p. 13.
17. O. P. Hopkins, Effect of war on American chemical trade, *Ind. Eng. Chem.*, 1918, **10**, 692.
18. Whaling statistics, 29 April 2013, http://luna.pos.to/whale/sta.html.

CHAPTER 9

Germany in a Fix

SAVE YOUR CHAMBER LYE

Early in 1915, German newspapers carried the following advertisement:

> The women of Germany are commanded to save their chamber lye, as it is very needful to the cause of the Fatherland in the manufacture of nitre, one of the ingredients of gunpowder. Wagons, barrels, and tanks will be sent to residences daily to collect and remove the same.[1]

The command was signed by German field marshal Paul von Hindenburg (1847–1934) who had recently led German armies to decisive victories against Russian armies on the Eastern Front. It formed part of a campaign to mobilise all available raw material resources for Germany's war effort.

The campaign began on 9 August 1914, just after the outbreak of war, when German Minister of War Erich von Falkenhayn asked industrialist Walter Rathenau (1867–1922) to organise the collection of the country's raw materials for the manufacture of explosives and other war supplies. Rathenau rapidly set up the War Raw Materials Board to inventory, recycle, and supply whatever materials could be collected. In addition, the board

The Chemists' War, 1914–1918
By Michael Freemantle
© Freemantle, 2015
Published by the Royal Society of Chemistry, www.rsc.org

instigated food rationing for the German civilian population. Nothing was to be wasted and nothing was to be excluded, not even fruit peel, women's hair, farmyard manure, and the contents of chamber pots.

Chamber lye, that is the urine collected in chamber pots, is rich in nitrogen compounds, not least urea. When urine is left to stand, the dissolved urea reacts with water and breaks down into ammonia and carbon dioxide in a process known as hydrolysis. The process is catalysed by a bacterial enzyme called urease. The board aimed to supply the chamber lye to industry so that its dissolved nitrogen compounds could be converted into ammonia which could be used to produce potassium nitrate (also known as nitre). The nitrate would then be mixed with charcoal and sulfur to produce gunpowder.

The use of natural resources and waste to produce ammonia and other nitrogen compounds was well known. In the third part of a 1916 treatise on coal tar and ammonia that extended to over 1000 pages, George Lunge, a chemistry professor at the Federal Technical University, Zurich, described the production of ammonia from sewage, guano (bird excrement), bones, horn, leather, wool, hair, peat, shale, coal, urine, coal and other sources.[2] In one macabre example, he referred to the preparation of ammonium carbonate in the early 18th century by the distillation of the skull of a man who had died on the gallows.

Various methods for converting these forms of fixed nitrogen into nitrates were well established. In 1862, for example, the American Joseph LeConte (1823–1901), a professor of chemistry and geology at South Carolina College, outlined a process for converting farm animal urine into nitrates that could be used to fertilise the soil. The process was called the Swiss method because it was practised on small farms in Switzerland. It involved absorbing the urine in pits of porous sand. The urine "becomes nitrified, and is fit for leaching in about two years," he noted in his description of the method.[3]

It is unlikely that the War Raw Materials Board considered this lengthy nitrification process for the manufacture of nitrates. In all probability, the board planned to convert the chamber lye ammonia into nitric acid by a method developed by German chemist and Nobel Prize winner Friedrich Wilhelm Ostwald. The process, patented in 1902, employs atmospheric oxygen to convert ammonia into nitrogen dioxide which is then dissolved in water to form nitric acid. Platinum is used as a catalyst to speed up the process (see Chapter 11).

Finally, the potassium nitrate needed to make gunpowder could be manufactured by the reaction of the nitric acid with a potassium compound such as potassium hydroxide. At the time, Germany manufactured a variety of potassium compounds from the potash that occurred naturally in abundance in Germany (see Chapter 17).

GERMANY'S DESPERATE PLIGHT

Whether Hindenburg's chamber lye directive "ever amounted to much, I am unable to say," observed Richard Zucker in a report on Germany's industrial position published in 1919.[1] At the time, Zucker was stationed at the United States 3rd Army's troop headquarters in Germany. His report, which referred to the Hindenburg advertisement, described how Germany had become acutely short of certain raw materials and its industrial plight desperate by the start of 1915.

During the latter half of the 19th century, the country had rapidly developed its chemical industries and immediately before the outbreak of the First World War in 1914, it was exporting pharmaceuticals, dyes, and other chemical products on a grand scale to many parts of the world, but this manufacture depended to a large extent on raw materials imported from other countries. For example, Germany had no significant copper, nickel or tin deposits.

The country did have rich coal deposits, however, and much of its industrial expansion in the second half of the 19th century and early 1900s can be attributed to the exploitation of these deposits. For example, during the 1880s, Germany produced around 47 million tons of coal each year.[4] By1913, production had risen to 191 million tons. Much of it was consumed by state railways and the blast furnaces that produced iron and steel.

To meet the burgeoning demands of heavy industry, Germany also imported coal and coal tar from Britain before the war. At the time, Britain produced more coal than any other country in the world, but unlike Germany, Britain did not exploit coal tar to any significant extent to provide the chemicals needed for the manufacture of synthetic organic chemical products such as dyes and pharmaceuticals. Britain imported these products from Germany as did many other countries.

Although Germany was rich in coal and had the manpower and scientific and technical expertise to convert coal tar into profitable chemical products, it soon began to experience shortages of nitrogen-containing raw materials after the war broke out. Like every other country in the world, Germany had an abundance of atmospheric nitrogen but this type of nitrogen is inert and of limited use unless converted into nitrogen-containing compounds in a process known as nitrogen fixation. Water-soluble compounds containing fixed nitrogen, such as ammonium sulfate, potassium nitrate and sodium nitrate were widely used in agriculture as fertilisers. Nitrates were also required as raw materials to make explosives such as gunpowder and TNT.

Coal contains significant amounts of fixed nitrogen. In the decades leading up to the First World War, ammonium sulfate was produced as a by-product in large quantities when coal was distilled to make coke, coal gas and coal tar (see Chapter 5). At the end of the 19th century Great Britain alone was distilling well over 13 million tons of coal to produce some 700 000 tons of coal tar and 200 000 tons of ammonium sulfate, Lunge noted in the first part of his tome on coal tar and ammonia.[5]

Germany could not match this level of production. In 1913, the country consumed 225 000 tons of fixed nitrogen, 85–90% as fertilisers and 10–15% in industrial uses, according to British industrial chemists Bryan Reuben (1934–2012) and Michael Burstall.[6] Production of by-product ammonium sulfate from coal accounted for 119 000 tons of the total. The country imported most of the remaining 106 000 tons as Chile saltpetre, a sodium nitrate mineral found extensively in deposits in Chile and Peru. The nitrate was used as a fertilisers and also as a raw material for the manufacture of explosives.

The outbreak of war and the British Royal Navy blockade put a stop not only to German exports of pharmaceuticals, dyes and other valuable chemicals, but also to imports of coal and coal tar from Britain and Chile saltpetre from South America. The impact on German industry was dramatic. For example, production of by-product ammonium sulfate dropped to 54 000 tons by 1918, exactly one half of the amount produced in the year before the war. The Germans captured significant quantities of Chile salt-petre when they invaded Belgium in 1914 but then, as the blockade began to bite, stockpiles of this form of fixed nitrogen rapidly diminished.

The blockade and cessation of foreign trade created an immense dilemma for Germany. The country had envisaged a short sharp victorious campaign in France. The plan, devised by German field marshal Count Alfred von Schlieffen (1833–1913) in 1905, anticipated thrusting speedily through Belgium into northern France and then cutting off Paris from the sea within a few weeks. Victory would release troops to fight Russia on the Eastern Front.

Initially, the plan was successful but then, in early September 1914, French armies assisted by the British Expeditionary Force halted the Germans after they crossed the River Marne east of Paris. The Germans fell back north across the River Aisne where they dug in. Static trench warfare had begun and was to last until 1918. The Battle of the Marne proved to be a critical moment in the war following which, it can be argued, Germany would find it impossible to win.

With its widespread belief that the conflict would be over quickly, Germany had entered the war with limited stocks of food and just enough stocks of ammunition for an intensive campaign of a few months. The advent of a style of static warfare that relied on repeated artillery bombardments of enemy trenches meant that the country soon had to replenish its stocks of food and ammunition, but that required supplies of fixed nitrogen for use as fertilisers and for the manufacture of explosives. Within months after the start of the war Germany faced several shortages of both fertilisers and explosives.

The year 1915, when von Hindenburg issued his command to the women of the Fatherland, proved to be the low point in the manufacture of fixed nitrogen. Production by various methods totalled just 96 000 tons. Bearing in mind that Germany had consumed 225 000 tons in 1913, this relatively small amount created an immense challenge for the Germans especially as increasing quantities were now needed for the manufacture of explosives.

The country began to exploit every available source of fixed nitrogen. First it boosted the production of fixed nitrogen by the cyanamide process. Calcium cyanamide, also known as lime nitrogen, is produced by heating calcium oxide, that is lime, and coke in an atmosphere of nitrogen. The cyanamide reacts with water to form ammonia. Production of fixed nitrogen in Germany by this method increased from 10 000 tons in 1913 to 36 000 tons in 1918.

The Haber process for fixing nitrogen also became increasingly important during the war. In 1908, German chemist Fritz Haber (1868–1934) showed that ammonia could be synthesized from nitrogen extracted from air and hydrogen prepared by the electrolysis of water or by passing steam over red-hot coke. The nitrogen and hydrogen were heated in the presence of iron which acted as a catalyst to speed up the reaction. Carl Bosch, an industrial chemist working for the German chemical firm BASF, subsequently designed a reactor that allowed the process to be carried out more efficiently at high pressures and temperatures. The first industrial Haber ammonia synthesis plant came on stream in September 1913 and by the end of the year 1000 tons of fixed nitrogen in the form of ammonium sulfate had been produced. The sulfate was initially used as a fertiliser but then, in 1914, the process was employed to generate the ammonia needed to make nitric acid for the manufacture of explosives. Production was slow at first with only 6000 tons and 12 000 tons of fixed nitrogen being produced in 1914 and 1915, respectively. Production then accelerated. By 1916, the Haber process accounted for 30% of the country's fixed nitrogen production and by 1918, the process produced 95 000 tons, which was more than half the total fixed nitrogen production.

Thus, as the war progressed, Germany was able to supplement its rapidly diminishing stocks of Chilean nitrates and other forms of fixed nitrogen, notably ammonium sulfate generated as a by-product from coal, with nitrogen compounds manufactured by the cyanamide process and most importantly the Haber and Ostwald processes.

TO STARVE OR TO SHOOT

But for this effort to produce nitrogen compounds, "Germany would have run out of food and explosives in 1916 and the war would have ended," concluded Reuben and Burstall in their book on the chemical economy.[6]

Haber himself put the dilemma even more bluntly in a lecture before German officers in 1920.[7] "The uncompromising alternative was to starve or to shoot," he said. He was talking specifically about the glycerine and nitric acid needed to make nitroglycerine and other explosives.

Glycerine was produced from animal and plant fats and oils, but, as Haber pointed out, Germany never had enough pig fat, goose fat, mutton grease, butter and other fats and oils to meet all the country's needs. The German war effort would have been severely compromised had it not been for the pioneering biochemist Carl Neuberg (1877–1956) and other German scientists who early on in the war developed industrial processes for preparing glycerine by fermenting sugar solutions.

Haber went on to say that the supply of Chile saltpetre in the country at the start of the war was not only critical but also that the stocks of the mineral captured in Belgium had little impact on the situation. In the autumn of 1914, "the absolute necessity of ending the war in the spring of 1915 was known to every expert," he added. As a consequence, the high pressure synthesis of ammonia from the nitrogen in air and the production of lime nitrogen during the war grew to enormous importance."

Germany's war-time problems were compounded by the army, industry, and farms all competing for male labour in the first two years of the war. In December 1916, the government introduced a law to mobilise all men between the ages of 17 and 60 into the army or the industries that processed chemicals and metals and produced guns and ammunition. On top of all this, the country experienced poor harvests, particularly in 1917 when the potato crop failed following a poor winter. The winter became known as the "turnip winter" because barely edible turnips normally fed to farm animals were used as a substitute for potatoes and meat.

And so, even with rationing imposed by the War Raw Materials Board, many of its population faced severe food shortages. In the years leading up to the war, Germany had imported around 30% of its food, most notably grain from Russia. By the summer of 1916, the average weekly consumption of meat per head of population had dropped from just over 1.1 kg to less than 0.5 kg.[8] The consumption of flour and fats also fell significantly as did the quality of foods. "Miserable coffee substitutes were made from acorns, horse chestnuts, and carrots," noted Zucker in his report on Germany's industrial position.[1] "Dehydrated vegetables and potato flour played an important part in the feeding of the nation," he added.

By the autumn of 1918, Germany had become almost self-sufficient in fixed nitrogen but in the end, German industry could not, "deliver both guns and butter," according to Roy MacLeod, a historian at the University of Sydney, Australia.[9] American author Daniel Charles called Fritz Haber "the patron saint of guns and butter," but notes that an estimated 750 000 Germans perished from the effects of hunger over the course of the war.[10]

Germany put a premium on guns rather than butter. The policy thus led to extensive malnutrition, starvation, loss of morale, and loss of life on the home front. If it had focused its industrial energy and resources more on butter, it is unlikely that the country could have manufactured the ammunition in the quantities needed to sustain the war of attrition that began in the trenches following the Battle of Marne in September 1914 and continued until 1918.

REFERENCES

1. R. D. Zucker, Germany's industrial position, *Ind. Eng. Chem.*, 1919, **11**, 692.
2. G. Lunge, *Coal-tar and Ammonia*, Part III, 5th edn, Van Nostrand, New York, 1916, Chapter 13.
3. J. LeConte, *Instructions for the Manufacture of Saltpetre*, Charles P. Pelham, State Printer, Columbia, SC, 1862, p. 9.
4. V. Berghahn, Demographic growth, industrialization and social change, in *Nineteenth Century Germany: Politics, Culture and Society 1780–1918*, ed. J. Breuilly (ed), Arnold, London, 2001, p. 189.
5. G. Lunge, *Coal-tar and Ammonia*, Part I, 4th Edn, Gurney and Jackson, London, 1909, p. 13.
6. B. G. Reuben and M. L. Burstall, *The Chemical Economy: A Guide to the Technology and Economics of the Chemical Industry*, Longman, 1973, p. 18.
7. F. Haber, Chemistry in War, *J. Chem. Educ.*, 1945, **22**, 526.

8. G. Jukes, *The First World War: The Eastern Front 1914–1918*, Osprey, Oxford, 2002, p. 82.

9. R. M. MacLeod, Chemistry for King and Kaiser: Revisiting Chemical Enterprise and the European War, in *Determinants in the Evolution of the European Chemical Industry, 1900–1939*, ed. A. S. Travis, H. G. Schröter, E. Homburg and P. J. T. Morris, Kluwer Academic Publishers, Dordrecht, The Netherlands, 1998, p. 44.

10. D. Charles, *Between Genius and Genocide: the Tragedy of Fritz Haber, Father of Chemical Warfare*, Pimlico, London, 2006, p. xiii and p. 175.

CHAPTER 10

May Sybil Leslie

AN UNSUNG CHAMPION

May Sybil Leslie is one of the unsung champions of World War I chemistry. Details of her life and work were uncovered by chemistry professor Geoff Rayner-Canham and laboratory instructor Marelene Rayner-Canham at Sir Wilfred Grenfell College, Memorial University of Newfoundland, Canada, while carrying out a study of pioneer women in nuclear science.[1] In a subsequent article, the Rayner-Canhams described Leslie as a chemist of some repute who made a major impact in industry and academe in the early art of the twentieth century.[2] "Her main claim to fame is the industrial chemistry research that she did during World War I," they noted.

Leslie was born on 14 August 1887 in Woodlesford, a village near Leeds in Yorkshire, England. In 1908, she graduated with a first class honours degree in chemistry at Leeds University. She then carried out research on physical chemistry at the university under the supervision of Harry Medforth Dawson (1876–1939), professor of physical chemistry.[3] Dawson had been "attracted to the study of chemistry" at Leeds by the teaching of chemistry professor Arthur Smithells (1860–1939). Like many British chemists of the time, Dawson and Smithells both carried out postgraduate research in Germany (see Chapter 18).

The Chemists' War, 1914–1918
By Michael Freemantle
© Freemantle, 2015
Published by the Royal Society of Chemistry, www.rsc.org

The university awarded Leslie a Master of Science degree in 1909. The following year, she won a scholarship to spend two years working at the Radium Institute in Paris with Marie Curie who was by then a world famous physicist. Five years earlier, Curie had shared the Nobel Prize in Physics with her husband Pierre and Antoine Henri Becquerel for their work on radioactivity (see Chapter 4).

Leslie's research at the institute focused on the extraction of elements from ores containing thorium, a radioactive element. In 1911, she returned to England and carried out research at Victoria University of Manchester on the radioactivity of thorium and actinium, another radioactive element. In the same year, Curie was the sole winner of the Nobel Prize in Chemistry for the discovery of the radioactive elements polonium and radium (Figure 10.1).

Figure 10.1   May Sybil Leslie.

Leslie worked in Manchester with Ernest Rutherford (1871–1937), physics professor at the university, who had won the 1908 Nobel Prize in Chemistry "for his investigations into the disintegration of the elements, and the chemistry of radioactive substances." In 1912, Leslie left Manchester to teach science at a high school in West Hartlepool, a town in North East England. Two years later, she took up a post as assistant lecturer and demonstrator in chemistry at University College, Bangor, Wales. While teaching at the high school and the college, she also carried out research in collaboration with Dawson.

Her war work started in 1915 when the government hired her as a chemist to carry out research in a laboratory at His Majesty's Factory in Litherland, Liverpool. The factory occupied part of a site where the chemical company Messrs Brotherton distilled coal tar. Leslie investigated the chemistry involved in the manufacture of nitric acid and optimised the conditions for the process. The work enabled nitric acid to be produced on an industrial scale much more efficiently. Large quantities of the acid were required to manufacture the vast amounts of explosives needed by the munitions industry for the war effort. Leslie was promoted to chemist in charge of the laboratory in 1916. It was "a very high position for a woman at that time," the Raynham-Canhams remark.[2]

In 1918, Leslie was transferred to His Majesty's Factory in Penrhyndeudraeth, a village in Wales. In the same year, she was awarded a doctorate by the University of Leeds in recognition of her research on radioactivity and her contributions to the large scale manufacture of nitric acid and explosives. Her work on high explosives "contributed effectively towards the solution of urgent technical problems," commented Dawson in his Chemical Society obituary of her.[4] Nevertheless, when male chemists returned from the war, Leslie was relieved of her government appointment. She returned to the University of Leeds where she became a demonstrator in the Department of Chemistry.

Five years later, Leslie married Alfred Hamilton Burr, a lecturer in chemistry at the Royal Technical College, Salford, which is now the University of Salford, Greater Manchester, England. Burr had also worked at the Litherland factory during World War I.

May Sybil Leslie Burr, as she now was, was promoted to lecturer in physical chemistry at Leeds University in 1928. During her career, she published a number of papers on her research, some with Dawson, and also a textbook on inorganic chemistry.[5]

She died on 3 July 1937, a month short of her 50th birthday. "No cause of death was recorded, but it was quite possibly radiation-related, considering her young age and her experiences in Paris," according to the Rayner-Canhams.[2]

REFERENCES

1. M. F. Rayner-Canham and G. W. Rayner-Canham, Pioneer women in nuclear science, *Am. J. Phys.*, 1990, **58**, 1036.
2. G. Rayner-Canham and M. Rayner-Canham, A chemist of some repute, *Chemistry in Britain*, 1993, **29**, 206.
3. R. Whytlaw Gray and G. F. Smith, Harry Medforth Dawson: 1876–1939, *J. Chem. Soc.*, 1940, 564.
4. H. M. Dawson, May Sybil Burr: 1887–1937, *J. Chem. Soc.*, 1938, 151.
5. *A Textbook of Inorganic Chemistry*, ed. M. S. Burr and J. Newton Friend, vol. 3, part 1, Charles Griffin, London, 1926.

CHAPTER 11

An Element of War

PLATINUM PLEDGE

In 1917, the Daughters of the American Revolution (DAR), a national society of women dedicated to promoting patriotism and preserving American history, pledged to refuse to purchase or accept as gifts for the duration of the war, jewellery and other articles made in whole or in part of platinum.[1] The pledge, made soon after the United States declared war on Germany, aimed to ensure that all possible supplies of the precious metal "shall be available for employment where they can do the greatest good in the service" of the country. At the same time, the American Chemical Society (ACS) launched a "movement for the conservation of platinum" at a meeting in Kansas City.[2] The National Academy of Sciences, based in Washington, DC, like DAR and ACS, made a similar appeal.

A few months later, in August 1917, G. Shaw Scott, secretary of the British Institute of Metals, wrote to ACS secretary Charles L. Parsons proclaiming that as platinum played such a vital part in the war it was "nothing short of a crime to allow its use for purposes of personal adornment."[3] The letter invited the society to cooperate with the institute in efforts to conserve platinum. Within a few months, a Women's National League for the Conservation of Platinum had been established in the United States.

The Chemists' War, 1914–1918
By Michael Freemantle
© Freemantle, 2015
Published by the Royal Society of Chemistry, www.rsc.org

The country's government subsequently ordered that all crude and unworked platinum in the hands of importers, jobbers, and wholesalers be commandeered for the war effort.

Platinum, like gold and silver, is a chemical element known as a precious metal because of its rarity. Much of the platinum used in the early 20th century originated in mineral deposits in the Ural Mountains of Russia. Although rare, the refined metal was widely available—at a price—throughout Europe and the United States. Between 1911 and 1915, Russia exported a total of 551 400 ounces (almost 15 000 kg) of platinum to France, 71 624 ounces (just over 2000 kg) to England, and 178 534 ounces (about 5000 kg) to Germany.[4] It was used not just for making jewellery but also in dentistry and for making laboratory equipment and a variety of other items of commercial and industrial importance.

It was critical for the war effort on all fronts of the Great War for a variety of reasons. The metal was used, for example, to make pyrometers—instruments that measured high temperatures. The instruments were essential for the manufacture of the steel needed to make guns, battleships, and railways.

CATALYST FOR SULFURIC ACID PRODUCTION

Platinum played a crucial role in incorporating the chemical element nitrogen into many of the explosives required by the munitions industries of the belligerent nations. The high explosive trinitrotoluene (TNT), ammonium nitrate—a component of explosives such as amatol and ammonal, and the nitroglycerine and nitrocellulose used to make the World War I propellant cordite, all contained nitrogen. It was this element that gave these chemicals their powerful explosive properties.

Vast amounts of nitrogen-containing explosives were produced during the war (see Chapter 1). All these explosives required nitric acid as a source of nitrogen for their production. TNT, for instance, was produced by treating toluene, a chemical obtained by distilling coal tar or petroleum, with a mixture of concentrated nitric and sulfuric acids. Nitric acid itself was manufactured by the reaction of sodium nitrate and concentrated sulfuric acid.

Much of the sulfuric acid needed for the production of nitric acid and explosives such as TNT was manufactured by what is known as the contact process, a process that had been developed and used extensively in Germany. The raw materials for the process were air, water, and sulfur or a sulfur-containing mineral such as iron pyrite. The platinum was used as a catalyst to speed up a key chemical reaction in the process.

An alternative industrial process, known as the lead chamber process, was also employed to manufacture the acid. The process used the same sulfur raw materials but did not require platinum as a catalyst. However, although widely used, especially in Britain before the war, it was less economic than the contact process. In 1915, Britain began to turn its attention to the contact process for production of the acid.[5]

There can be little doubt that without efforts to conserve platinum and ensure that sufficient quantities of the precious metal were available for the contact process, the Allies would have struggled to manufacture the amounts of sulfuric acid required for the production of nitric acid and explosives.

CATALYST FOR NITRIC ACID PRODUCTION

Germany faced an even greater problem when it came to manufacturing the nitric acid needed for explosives. Before the war Germany, like Britain and other countries, had produced the acid by the reaction of sulfuric acid and sodium nitrate. The latter was imported as a raw material from Chile and other South American countries. Known as Chile saltpetre, it was used not only to make nitric acid and therefore explosives, but also as a nitrogen-containing soil fertilizer.

In 1913, Germany imported 750 000 tons of the mineral from Chile and by the start of the war had a stockpile of some half a million tons. However, the British Royal Navy blockade of the country imposed at the outbreak of the war in August 1914 blocked further imports of the mineral. As existing stocks rapidly became depleted, Germany faced the imminent threat of being starved not only of the explosives demanded by the munitions industries but also of the fertilisers required for food production (see Chapter 9).

German chemists therefore turned to another source of the nitrogen needed to make nitric acid and nitrogen-containing explosives. That source was air, which contains around 78% of the element. The conversion of the nitrogen in air into nitric acid involved two chemical processes, the second of which, like the contact process for making sulfuric acid, relied on the use of platinum as a catalyst.

The first process exploited the nitrogen in air as a raw material to make ammonia. The second process converted the ammonia into nitric acid.

At the time of the First World War, crude ammonia was commonly obtained as a by-product in the production of coke and coal gas from coal. However, impurities in the by-product reduced its effectiveness for the manufacture of nitric acid. In 1908, the German chemist Fritz Haber had shown that a much purer form of ammonia could be synthesised by the reaction of nitrogen extracted from air, and hydrogen—an element produced by the electrolysis of water or by passing steam over white-hot coke. German industrial chemist Carl Bosch and colleagues subsequently developed an industrial-scale version of the Haber process. The German chemical firm BASF started manufacturing ammonia using the Haber-Bosch process in 1913. Initially, the firm used the ammonia to make ammonium sulfate, a soil fertiliser. Consequently, when war broke out, Germany was in a good position to exploit the process for the production of the ammonia in the quantities needed to make nitric acid and explosives.

The second process, patented by German chemist and Nobel Prize winner Friedrich Wilhelm Ostwald in 1902, employed platinum as a catalyst to speed up the chemical reaction between ammonia and atmospheric oxygen. The Ostwald process, as it is known, leads to the formation of nitrogen dioxide, a brown gas that dissolves in water to form nitric acid.

If it had not been for the ingenuity of German chemists such as Haber, Bosch and Ostwald and the availability of the precious metal platinum, the war may not have lasted for more than a year or so and it is quite possible that Germany might have faced defeat by Christmas 1914 or early 1915, as had widely been predicted in Britain at the outbreak of war.

Without the Haber-Bosch process, "Germany could not have made the nitric acid required for her explosives programme, nor

Figure 11.1 Friedrich Wilhelm Ostwald (courtesy: Archiv der Max-Planck-Gesellschaft, Berlin-Dahlem).

obtained fertilisers for food production after our blockade, and it is probable that she could not have continued the war after 1916," concluded British chemist Harold Hartley (1878–1972) in 1919.[6] Hartley, a lecturer in chemistry at Oxford University, served in the British Army and by the end of the war had been promoted to the rank of brigadier general and Controller of the Chemical Warfare Department at the British Ministry of Munitions.

The war could even have been shorter according to Leo Baekeland (1863–1944), the Belgian-American industrial chemist who invented and patented the plastic Bakelite in 1907. Speaking to chemists at a joint meeting of ACS and the American Electrochemical Society in New York in 1915, he observed that Germany would have been "hopelessly paralysed" had it not been for the development of chemical processes that relied on platinum for the manufacture of nitric acid from air.[7]

REFERENCES

1. C. L. Parsons, Platinum in jewelry, *Ind. Eng. Chem.*, 1917, **9**, 622.
2. Anon., The platinum situation, *Ind. Eng. Chem.*, 1917, **9**, 544.
3. S. Scott, British control of platinum, *Ind. Eng. Chem.*, 1917, **9**, 731.
4. A. R. Merz, Russia's production of platinum, *Ind. Eng. Chem.*, 1918, **10**, 920.
5. R. MacLeod, The chemists go to war: mobilisation of civilian chemists and the British war effort, 1914–1918, *Ann. Sci.*, 1993, **50**, 455.
6. H. Hartley, Military importance of the German chemical industry, *Ind. Eng. Chem.*, 1921, **13**, 284.
7. L. H. Baekeland, The Naval Consulting Board of the United States, *Ind. Eng. Chem.*, 1916, **8**, 67.

CHAPTER 12

Fritz Haber: Revered and Reviled

A HEROIC AND TRAGIC FIGURE

One day in the summer of 1933, two of the most important chemists of the First World War met in Switzerland. They were the German Fritz Haber and Russian-born Chaim Weizmann who had become a British subject. Early on in the war, Weizmann had famously developed an industrial fermentation process for the manufacture of the acetone that Britain needed to make the propellant cordite (see Chapter 7). An ardent supporter of the Zionist movement which aimed to establish a homeland for the Jews in Palestine, Weizmann would later become the first president of the State of Israel.

At the time of the meeting, Weizmann was staying at Zermatt with his wife, Vera (1881–1966), and their son Michael (1916–1942). Vera, like her husband, was a Zionist activist. Michael was to become a pilot in the British Royal Air Force during the Second World War and was killed when his plane was shot down off the coast of France in February 1942. As the meeting progressed, Chaim invited Haber to dine with his family that evening.

During the dinner, according to Weizmann,[1] Haber broke into an eloquent tirade: "Dr. Weizmann, I was one of the mightiest men in Germany. I was more than a great army commander, more than a captain of industry. I was the founder of industries;

The Chemists' War, 1914–1918
By Michael Freemantle
© Freemantle, 2015
Published by the Royal Society of Chemistry, www.rsc.org

my work was essential for the economic and military expansion of Germany. All doors were open to me."

But by the summer of 1933, the doors had been slammed firmly shut in his face. He was no longer director of the Kaiser Wilhelm Institute for Physical Chemistry and Electrochemistry at Dahlem, Berlin, a position he had held since the institute was founded in 1911. Haber, a German war veteran, a German patriot, and the leading German chemist of the First World War, was now *persona non grata* in his homeland. In January 1933, Adolf Hitler (1889–1945) had been appointed chancellor of Germany. Within a few months Haber, who was born into a Jewish family, resigned from the institute having been forced to fire his Jewish colleagues at the institute. Haber left Berlin and Germany for good in August that year to escape the Nazi's anti-Semitism.

At the dinner with the Weizmanns, Haber went on to praise the Weizmanns' efforts in the Zionist movement. He then added

Figure 12.1   Fritz Haber (Royal Society of Chemistry Library).

sadly: "At the end of my life I find myself a bankrupt." The man who had once been rich and famous in Germany had now been stripped of most of his money, his position at the institute in Berlin, and his honours. He was also homeless, asthmatic and suffering from heart problems. He died in Basel, Switzerland, "a broken, confused, lonely and bitter man," remarks David Cahan, a history professor at the University of Nebraska–Lincoln, United States.[2]

Yet Haber was arguably the most important chemist of the 20th century, not just in Germany but in the whole world, and for several reasons. He is most widely known for the process that is named after him. Well before the First World War, he developed a workable process for converting chemically-free nitrogen in the atmosphere into chemically-fixed nitrogen in the form of ammonia. German industrial chemist Carl Bosch (1874–1940) and colleagues subsequently developed his nitrogen fixation process and used the ammonia to manufacture nitrogenous fertilizers in vast quantities for use in agriculture.

Haber had effectively found a way of making "bread from air." The Haber–Bosch ammonia synthesis process, as it became known, helped feed millions of people around the world and, according to some commentators, was largely responsible for averting famine and saving the world from starvation.[3,4] It detonated the population explosion in the 20th century, observed Vaclav Smil, a Czech-Canadian scientist and policy analyst at the University of Manitoba, Canada. Writing in 1999, Smil suggested that "the world's population could not have grown from 1.6 billion in 1900 to today's six billion," if it had not been for the process.[5] Without the process, "almost two-fifths of the world's population would not be here" he explained.

The ammonia synthesis process also played a critical role in the war. In the same way that Britain would have found it difficult to sustain its war effort for four years without the acetone produced by the Weizmann process (see Chapter 7), Germany could not have continued fighting the war beyond a few months without the Haber–Bosch process that enabled it to continue manufacturing nitrogen-containing explosives such as trinitrotoluene, nitroglycerine, and nitrocellulose (see Chapter 9). The process therefore facilitated the production of not just "bread from air" but also "explosives from air."

Haber's nitrogen fixation process was not his only contribution to the applied chemistry of the 20th century. He is also well known and much vilified for spearheading Germany's development of lethal chemical weapons and overseeing their introduction on the Western Front in April 1915 and their continued use throughout the war. The "Father of Modern Chemical Warfare," as Haber became known, considered chemical warfare to be humane (see Chapter 13). He was not alone in this view. Munitions and war experts in both the Central Powers and the Allied Powers shared this "perverse-sounding notion" notes Bretislav Friedrich, physics professor at the Fritz Haber Institute of the Max Planck Society in Berlin.[6] Haber "regarded chemical weapons as a means to break the stalemate of trench warfare, shorten the war, and therefore preclude the slaughter of millions by artillery and machine gun fire," Friedrich remarks.

Haber's immense contributions to Germany's manufacture of explosives and to chemical warfare unquestionably had a monumental impact on the nature, course and duration of the war. Yet curiously, Haber receives scant if any attention in many of the more popular books on the history of World War I. One can search in vain for his name in the pages and indexes of such books.

HABER'S EARLY LIFE

Fritz Haber was born on 9 December 1868 in Breslau, an industrial city in the Kingdom of Prussia. The kingdom became part of the German Empire on the unification of its constituent territories three years later. The city is now called Wroclaw and is one of the largest cities in Poland.

Fritz was the son of Siegfried Haber, a prosperous Jewish chemical and dye manufacturer and merchant, and his wife Paula who was also his cousin. The birth was difficult and painful. Paula did not recover and died three weeks later.[7] In 1886, Fritz enrolled as a student at the University of Berlin and then after one year moved to the University of Heidelberg where one of his teachers was the ageing German chemist Robert Bunsen who had invented his famous burner some three decades earlier. After a spell in the army, Haber resumed his studies at the Technical Institute of Charlottenburg near Berlin. During this period, he became fascinated with organic chemistry. After completing his final examination in 1891, he spent short periods working in a cellulose factory, a distillery, and a chemical plant before spending one semester studying chemical technology at what is now ETH Zurich (*Eidgenössische Technische Hochschule Zürich*, the Swiss Federal Institute of Technology in Zurich) where coal tar expert George Lunge was a chemistry professor (see Chapter 9).

In 1892, Fritz converted to Christianity for reasons that are unclear,[8] and in the same year moved to the University of Jena in eastern Germany where he continued studying chemistry and worked as a laboratory assistant. He also carried out research in organic chemistry under German chemistry professor Ludwig Knorr (1859–1921) who had synthesised antipyrine, an analgesic and antipyretic drug, while at the University of Erlangen. It was the first synthetic drug.

Haber found academic life attractive at Jena and remained there for two years before taking up a junior appointment at the *Technische Hochschule Karlsruhe* (Technical Institute of Karlsruhe) in southwest Germany (Figure 12.2). The institute received substantial funding from BASF (*Badische Anilin- und Soda-Fabrik*, Baden Aniline and Soda Factory), Germany's largest chemical company, which was located at Ludwigshafen, also in southwest Germany.

With his teaching duties and plenty of opportunities to carry out research, "he had gained the start he desired, and moreover he counted himself fortunate to be associated with an institution that stood in such close relation with industry," remarked British chemist Joseph Coates in "The Haber Memorial Lecture" delivered before the Chemical Society in London on 29 April 1937.[9] Coates worked with Haber in Karlsruhe and subsequently became professor of chemistry at University College, Swansea, Wales.

Haber began to investigate the thermal decomposition and combustion of hydrocarbons and soon became attracted to the field of physical chemistry. Among his many interests at the time were the electrolysis of compounds such as nitrobenzene and hydrochloric acid, the electrochemistry of iron, and the corrosion of underground water pipes and gas mains. He also wrote a book on the thermodynamics of gas reactions that was published in 1905.[10]

Figure 12.2   Haber in a laboratory at Karlsruhe (courtesy: Archiv der Max-Planck-Gesellschaft, Berlin-Dahlem).

By then, Haber had been married for four years to Clara Immerwahr (1870–1915) who had been born into a Jewish family near Breslau. He first met her when she was twenty-one and he was a chemistry student. In 1900 she was awarded a doctorate in chemistry at the University of Breslau, the first woman to receive such a degree from a German university. Her academic adviser was German physical chemist Richard Abegg (1869–1910), a chemistry professor at the university and a friend of Haber. Clara stayed on at the university as Abegg's laboratory assistant after gaining her doctorate. In 1901, Haber and Immerwahr resumed their acquaintance when they both attended a meeting of the German Electrochemical Society at Freiburg, a city near Karlsruhe. Haber proposed to Immerwahr at Freiburg, she accepted and within a few months they were married and living together in Karlsruhe.

Figure 12.3   Clara Immerwahr (courtesy: Archiv der Max-Planck-Gesellschaft, Berlin-Dahlem).

Haber was appointed Professor of Physical Chemistry and Electrochemistry and director of the Karlsruhe institute in 1906. It was the year that German physical chemist Walther Nernst (1864–1941), who had just been appointed professor of chemistry at the University of Berlin, showed that ammonia could be synthesised from its elements, nitrogen and hydrogen, at high pressure. Nernst later went on to win the Nobel Prize in Chemistry in 1921 "in recognition of his work in thermochemistry."

SYNTHESIS OF AMMONIA

Ammonia has a long history. English clergyman and chemist Joseph Priestley (1733–1804) isolated the gas in 1774. The first synthesis of ammonia from its elements is attributed to another

English chemist, Humphry Davy (1742–1786).[4] In 1807, he detected trace amounts of the gas following the electrolysis of distilled water. The hydrogen released from the water during the process had combined with nitrogen in the air to form the compound.

In 1904, the brothers Otto and Robert Margulies, owner and managing director respectively of the *Österreichische Chemische Werke* (Austrian Chemical Works) of Vienna asked Haber to investigate the possibility of synthesising ammonia from its elements. Haber pointed out that ammonia could be produced at low cost as a by-product in the manufacture of coke (see Chapters 5 and 9). Even so, the idea intrigued him and so he decided to look into the problem.

One of the papers on the topic that caught his attention was entitled "The Decomposition of Ammonia by Heat."[11] It was written by two British chemists: William Ramsay (1852–1916), professor of chemistry at University College, Bristol, and Sydney Young (1857–1937), a lecturer and demonstrator in chemistry at the college. The two chemists noted that "it has long been known" that ammonia decomposed into nitrogen and hydrogen when passed through a red-hot porcelain tube or when it was exposed to "the electric spark for some time."

In one of their experiments, Ramsay and Young passed dry ammonia through an iron tube heated to about 800 °C. They noticed that "the issuing gas" of nitrogen and hydrogen always had a faint smell of ammonia and was sufficiently alkaline to turn a piece of red litmus paper blue. They concluded that the decomposition was "never complete" and the minute amount of ammonia present in "the issuing gas" was possibly due to the decomposition of traces of water by iron. The process, they suggested, led to the formation of a highly active type of hydrogen known as "nascent hydrogen" that combined with nitrogen to form ammonia.

The results indicated "the possibility of a direct synthesis of ammonia from the elements," Haber observed in a lecture on the subject.[12] The observation inspired Haber to attempt the synthesis of ammonia by passing nitrogen and hydrogen over an iron catalyst at high temperatures. He decided to carry out the experiment at atmospheric pressure "because of the simpler apparatus required," Coates noted.[9] But Haber was unable to obtain more than a trace of ammonia. However, he did establish that an equilibrium existed between nitrogen, hydrogen, and ammonia and that water was not involved in the process. He described some of the work in his book on the thermodynamics of gas reactions.[10]

"At that point it seemed to me, in 1905, useless to pursue the problem further," Haber remarked in his lecture. He gave up only to resume interest the following year when Nernst informed him in a letter that small amounts of ammonia could also be produced if the reaction was carried out at high pressures.

Haber set to work on the project again in 1908, this time with a young British physical chemist from the Channel Islands, Robert Le Rossignol (1884–1976), who had studied chemistry under Ramsay at University College, London. Ramsay had moved from Bristol to London in 1887. Le Rossignol graduated from the college in 1905 and the following year was made a Fellow of the London-based Chemical Society.

Haber and his young assistant built a laboratory-scale high pressure apparatus that worked well at high temperatures and pressures up to 200 atmospheres, but they still needed to find a catalyst that was highly active under these conditions. They tested metals such as chromium, manganese, iron, nickel, osmium and uranium. Osmium, uranium, and iron all proved active. The two chemists then built a small-scale pilot plant that operated at 200 atmospheres and 600 °C and was capable of producing a few hundred millilitres of liquid ammonia per hour "with very low energy expenditure," according to Coates.[9] Haber and Le Rossignol filed a patent for their process in August 1909.[13]

The performance of the plant was good enough to encourage BASF to take a closer look. "It was an exciting day in July 1909" when two of the company's representatives visited the institute in Karlsruhe to see a demonstration of the process," Coates remarked.[9] By the end of the day they were "deeply impressed and completely convinced." The two representatives were industrial chemist Carl Bosch and his co-worker, the German chemist Alwin Mittasch (1869–1953). They quickly began the development of an industrial-scale process although it proved to be a formidable task, especially as some of problems were totally unforeseen.

Figure 12.4   Carl Bosch (courtesy: Archiv der Max-Planck-Gesellschaft, Berlin-Dahlem).

Building an industrial-scale plant that was safe to operate at 600 °C and a pressure of 200 atmospheres was a particular challenge. The high pressure reactors constructed by Bosch's team were made of steel—an alloy of iron and carbon. Some of the first reactors they built exploded as they could not withstand the high pressures. Bosch, who had knowledge and experience of not only chemistry but also metallurgy, soon realised that hydrogen at high temperatures and pressures reacted with the carbon in the alloy. As a result, the steel became brittle and liable to fracture. He solved the problem by fitting a lining of soft low-carbon iron inside the reactor.

Bosch and his colleagues also had to find inexpensive ways of producing nitrogen and hydrogen with sufficient purity for the ammonia synthesis. Nitrogen occurs in abundance in the atmosphere. The element comprises about 78% by volume of the air around us. The standard procedure for manufacturing the element involved cooling air until it liquefied and then fractionally distilling the liquid.

Haber and le Rossignol electrolysed water to prepare the hydrogen they needed for their experiments. The BASF team, on the other hand, produced hydrogen from water gas, a mixture of carbon monoxide and hydrogen obtained by passing steam over a bed of white-hot coke. The hydrogen was separated by cooling the water gas to -205 °C at which temperature the carbon monoxide and any other gases in the mixture liquefied—apart from hydrogen which remained as a gas.

Mittasch tackled the problem of finding a suitable catalyst for the process. Osmium was used initially but although highly effective it was scarce and like uranium too expensive. After carrying out thousands of experiments with a wide variety of catalysts and catalyst mixtures, they eventually settled on pure iron. Aluminium, calcium, and potassium oxides were added to the iron to boost its catalytic activity.

Haber had no share in the technical development of the BASF process, Coates pointed out in his 1937 memorial lecture to the Chemical Society.[9] The translation of his process to the industrial scale "under the leadership of Bosch stands out as perhaps the most difficult and brilliant feat of chemical engineering ever achieved," he remarked.

By 1911, BASF had started constructing an ammonia synthesis plant at Oppau, near Ludwigshafen. Production began in September 1913 primarily for the purpose of making nitrogeneous fertilizers. After the outbreak of war, ammonia became increasingly important in Germany for the manufacture of the nitric acid needed to make explosives (see Chapters 9 and 11). Coates, like many others, concluded that Haber saved Germany from premature defeat with his discovery of the process for nitrogen fixation.

In 1911, Haber accepted an invitation to become the founding director of the Kaiser Wilhelm Institute for Physical Chemistry and Electrochemistry at Dahlem, Berlin, a position he held until 1933 when Hitler and the Nazis came to power. Haber persuaded his close friend Richard Willstätter, who was professor of chemistry at the University of Berlin, to join him in Dahlem at the neighbouring Kaiser Wilhelm Institute for Chemistry which had also been founded in 1911 (Figure 12.5). Willstätter, who

Figure 12.5 Kaiser Wilhelm Institute for Chemistry, Dahlem, Berlin, around 1912 (courtesy: Archiv der Max-Planck-Gesellschaft, Berlin-Dahlem).

was born into a Jewish family in Karlsruhe, became director of the institute the following year.

Willstätter, an organic chemist, was renowned for his research on the structure and synthesis of natural products including haemoglobin, chlorophyll, and plant alkaloids such as cocaine and atropine. After the war, the Royal Swedish Academy of Sciences announced that he had been awarded the 1915 Nobel Prize in Chemistry "for his researches on plant pigments, especially chlorophyll," according to the Nobel citation.

At the same time, the academy named Haber as the winner of the 1918 Nobel Prize in Chemistry for his synthesis of ammonia from its elements. In his Nobel lecture on the topic, he paid tribute "with particular sincerity and gratitude" to the work of his "young friend and co-worker" Robert le Rossignol. Haber, always a generous man, subsequently recognised his young friend's contribution by rewarding him with some of his Nobel prize money. Le Rossignol also received a share of the substantial proceeds from the various patents that he and Haber had filed.

Le Rossignol left Karlsruhe before the war and joined *Auergesellschaft*, a German company that had developed the OSRAM light bulb. The bulb's filament was made from two metallic elements: OSmium and WolfRAM (also known as tungsten), hence the name. He was interned by the Germans in the autumn of 1914 but then released to continue his work for the company in March 1915. After the war, he returned to Britain to join the GEC research laboratory at Wembley, near London. He remained there working on electronics until the end of his career.

Bosch continued working for BASF during the war on the production of ammonia and on the development of processes for its conversion to the nitric acid required by the explosives industry. He was appointed managing director of the company in 1919. In 1931, Bosch and fellow German chemist Friedrich Bergius (1884–1949) were jointly awarded the Nobel Prize in Chemistry "in recognition of their contributions to the invention and development of chemical high pressure methods." Like Haber, Bergius was born in Breslau. He studied chemistry under both Nernst in Berlin and Haber in Karlsruhe. His most important research focused on the hydrogenation of coal and heavy oils under high pressure.

THE FATHER OF MODERN CHEMICAL WARFARE

By 1914, Haber was famous and revered as one of the most eminent scientists in the country. He embodied the "romantic, quasi-heroic aspect of German chemistry" that combined scientific advance and technological progress with national pride, commented his son Lutz Haber (1921–2004) in his book on chemical warfare.[14] ("Lutz" is a contraction of his first two names: Ludwig Fritz.)

The development of trench warfare following the Battle of the Marne in September 1914 had generated two urgent challenges for German chemists, chemical engineers, and the chemical industry. The first challenge was to ensure sufficient supplies of the chemicals needed to make munitions. The Haber–Bosch process played a central role in meeting this challenge. Secondly, the Germans decided to find some way of forcing the Allied troops out of their trenches. The use of irritant and lethal chemicals provided an answer. Once again, Haber led the way.

Fritz Haber was a Prussian and a loyal patriot. When the war started, he offered his services and those of his institute at Dahlem to the German War Ministry. Haber initially focused his efforts on generating supplies of nitrogen-containing compounds and other chemicals required for the war effort. He became, for example, a member of the "nitrate commission" which Germany had established towards the end of 1914.[15] At the same time, his institute and the neighbouring Kaiser Wilhelm Institute of Chemistry began carrying out research projects for the military even though many of the scientists in both institutes had been called up for military service.

There was evidence that as early as August 1914 both institutes were working on the development of poisonous chemical warfare agents for use in hand grenades, according to chemist Victor Lefebure (1891–1947) who was British Chemical Warfare Liaison Officer with the French during the war.[16] The chemicals included phosgene and cacodyl oxide. The Germans subsequently introduced phosgene in battle in December 1915 (see Chapter 13).

Cacodyl oxide, a highly toxic arsenic-containing organic compound also known as dimethylarsinous anhydride, was first prepared in a mixture with cacodyl (also named tetramethyldiarsine) by French pharmacist-chemist Louis Claude Cadet de Gassicourt (1731–1799) in 1757. The mixture became known as "Cadet's fuming arsenical liquid" or simply "Cadet's liquid." Although arsenic is not a metal, the components of the liquid mixture "played a vital role in the development of organometallic chemistry," according to Dietmar Seyferth, a chemistry professor at Massachusetts Institute of Technology, United States.[17] Bunsen was one of the first chemists to explore the chemistry of cacodyl compounds.

Dianisidine chlorosulfonate, a powdery organic chemical used in the manufacture of dyes, was one of the first irritant chemicals to be used by the Germans. When inhaled, it induces violent sneezing. The Germans attempted to boost the effectiveness of shrapnel shells by embedding the shrapnel balls in the powder. Their artillery fired some 3000 of the shells during a battle at Neuve Chapelle, France, on 27 October 1914 but they proved ineffective.

Xylyl bromide, a tear gas that the French had used in grenades against the Germans in the early months of the war, was equally ineffective as it was virtually impossible to generate sufficiently high concentrations of the gas when the grenades exploded. The Germans employed the chemical in gas shells on the Eastern Front in January 1915 but once again they had little impact.

The previous month, investigations of irritant chemicals in Haber's laboratory ended in tragedy when a test-tube of cacodyl chloride, an unstable compound, caught fire as a few drops of another organic chemical were being added to it. The fire ignited the bottle containing the organic chemical causing it to explode violently. The explosion filled the laboratory with dense clouds of arsenic compounds, blew off the hand of one of Haber's assistants and killed another, the up-and-coming physical chemist Otto Sackur (1880–1914) who, like Haber and Bergius, was born in Breslau. On hearing the explosion, Haber hastened to the laboratory and collapsed in shock at the scene. Haber and Willstätter both wept "uncontrollably" at Sackur's funeral.[18] Work on the project was immediately abandoned.

Figure 12.6   Kaiser Wilhelm Institute for Physical Chemistry and Electro-chemistry, Dahlem, Berlin, around 1912. The institute became the centre for German chemical warfare research during the war (courtesy: Archiv der Max-Planck-Gesellschaft, Berlin-Dahlem).

Artillery shells filled with a tear gas such as xylyl bromide only delivered small amounts of the gas when the shells burst. Haber was convinced that to be militarily effective, irritant chemicals needed to be delivered on a larger scale, for example by bombs fired from groups of trench mortars. He suggested to the German High Command that chlorine, which was widely available in the country, might be used as a chemical weapon and that clouds of the gas could be fired towards enemy trenches from cylinders—if the wind was in the right direction.

The High Command, led by Erich von Falkenhayn, warmed to his proposal and in January 1915 put him in charge of testing the gas as a weapon. The command did not consider the discharge of poisonous gas from cylinders to be a breach of the Hague Conventions of 1899 and 1907 which required signatories to abstain from the use of "projectiles" to deliver "asphyxiating or deleterious gases." Shells, grenades, trench mortar bombs were all projectiles, the Germans observed, but cylinders were not.

Haber recruited a team of 500 gas pioneers, including a number of scientists. They gathered at a site in Cologne and practised releasing chlorine from steel cylinders that had been filled with the gas under high pressure.[19] The soldier-scientist gas pioneers in the team included four future Nobel Prize winners: Haber; the German chemist Otto Hahn (1882–1964) who won the 1944 Nobel Prize in Chemistry for his research on nuclear fission; and two physicists—the German-born American James Franck (1882–1964) and the German Gustav Hertz (1887–1975) who shared the 1925 Nobel Prize in Physics for their research on energy transfer between molecules. Haber's close friend Willstätter, another Nobel Laureate, refused to work on the use of poison gases as chemical weapons but later helped to develop gas masks for the German army.

After ten weeks of trials and preparations, Falkenhayn authorised a chlorine cloud gas assault at Langemarck near Ypres in Belgium (see Chapter 13). By then, Haber's team had expanded into a Pioneer Regiment of some 1600 gas troops. Haber was in charge of the installation and release of the chlorine from over 5700 cylinders. After several postponements owing to unfavourable weather conditions, the gas was eventually discharged towards the Allied trenches on 22 April 1915.

Haber's associates on the project included chemist Hugo Stoltzenberg (1883–1974) who had worked with Heinrich Biltz as an assistant at the University of Breslau (see Chapter 15). Stoltzenberg was married to chemist Margarete Bergius, the sister of Nobel laureate Friedrich Bergius (see above). On the day of the gas attack, Stoltzenberg was in the team of gas troops who opened the cylinder valves to release the gas. One of the cylinders exploded blinding him in the left eye.[20]

Even so, the attack exceeded Haber's expectations and the Germans soon mounted further gas attacks near Ypres over the following weeks. At the end of April, Haber, who had now been promoted to the rank of captain, returned home to Dahlem to prepare for a trip to the Eastern Front where he was to supervise the first chlorine gas attack against the Russians. The Habers hosted a party on 1 May and then, that night, Haber and his wife Clara quarrelled. She had always been opposed to chemical

Figure 12.7   Haber in army uniform (courtesy: Archiv der Max-Planck-Gesellschaft, Berlin-Dahlem).

weapons and on several occasions had attempted to persuade her husband to stop his work on chemical warfare. In the early hours of 2 May, she took her husband's army pistol, walked out to their garden, and shot herself in the chest. She died in the arms of her 13-year-old son, Hermann (1902–1946). Obeying army orders, Fritz Haber left for the front the same day leaving Hermann alone to cope with his mother's suicide.

The successes of the first gas attacks on the Western Front convinced the Germans that the deployment of irritant and lethal chemicals could play a crucial role in overcoming the trench warfare stalemate and resuming a war of movement which they could win. They recognised, however, that chemical warfare would involve the use of not only chemical weapons but also defensive measures to protect against such weapons.

Haber was put in charge of gas defence. During the remaining months of 1915, Haber, Willstätter, and their colleagues at the Dahlem institutes worked on anti-gas measures. They developed the snout-canister gas masks that were subsequently produced in their millions to protect German soldiers against the Allied gas attacks. The troops breathed in filtered air through the canisters which contained a mixture of potassium carbonate, porous charcoal, and other substances. The mixture was designed to absorb or neutralise tear gases, chlorine, phosgene, and other poison gases.

In 1916, Germany established a Chemical Warfare Service with Haber as its chief. His institute in Dahlem was turned into a military establishment with every laboratory focusing on the development of either new chemical warfare agents or anti-gas measures. The institute conducted virtually all the country's chemical warfare research right through until the end of the war.

The most notorious of the institute's projects focused on mustard gas (see Chapter 14). The process for manufacturing the blister agent was developed by Wilhelm Steinkopf (1879–1949), the head of the institute's chemical weapons research team, and Wilhelm Lommel, a chemist at the Bayer chemical works in Leverkusen. Haber had initially dismissed the idea of using sulfur mustard, as it is also called, because he considered its toxicity to be too low. Its ability to blister skin was well known

Figure 12.8 Willstätter worked with Haber on anti-gas measures (courtesy:
Archiv der Max-Planck-Gesellschaft, Berlin-Dahlem).

but that was not thought to be significant at the time. When it
was first used against the British at Ypres in July 1917, the horrendous casualty toll surprised the Germans and only then did
they realise its potency as a chemical weapon. Haber subsequently regarded mustard gas as a "fabulous success."

Haber also introduced the concept of variegated shelling
during which two types of gas shells were fired simultaneously at
the enemy. One type of shell contained irritant chemicals,
known as sternutators, that caused sneezing or vomiting. They
were employed to force troops to remove their gas masks. The
troops would then be exposed to lethal gases such as phosgene
that were being released from the other gas shells.

On the personal front in 1917, Haber met and become engaged to Charlotte Nathan who was 20 years younger than him. It
was barely more than two years after Clara's suicide. The two
married in October that year and their first child, Eva, was born
in July the following year (Figure 12.9).

Figure 12.9   Charlotte Nathan and Fritz Haber (wearing the military uniform of a Prussian officer, including spiked helmet and sword) on their wedding day, 25 October 1917. They are pictured with Fritz's son Herman (on the left) in front of the Kaiser Wilhelm Memorial Church, Berlin (courtesy: Archiv der Max-Planck-Gesellschaft, Berlin-Dahlem).

AFTER THE WAR

Haber believed the war years to be "the greatest period of his life," according to Coates[9] Although he hated the wastage and suffering of war, he was also "autocratic and ruthless in his will to victory." Germany's defeat in 1918 therefore came as a tremendous shock and terrible tragedy for him. The outcome changed him and he never fully recovered from it. In particular, his efforts at his institute and on the battlefield during the war had exhausted him. By early 1919, Haber, now aged 50, was demoralised and suffering from depression and a heart condition.

He soon had another worry. Article 229 of the Treaty of Versailles, signed by the Allied Powers and Germany in June 1919, required that "persons guilty of criminal acts against the nationals" of the Allied Powers be brought before military tribunals. He suspected that he might be arrested as a war criminal. As a consequence, Haber sent Charlotte, Hermann, and Eva to Switzerland and followed them there in August, but he was not charged with any criminal acts and so returned to Dahlem later that year.

On 2 June 1920, Haber delivered his Nobel lecture in Stockholm. The lecture was entitled: "The synthesis of ammonia from its elements."[12] Willstätter presented his Nobel lecture "On plant pigments" the following day. His research on these pigments did not contribute to the German war effort and so understandably he did not refer to the war or his activities during the war in his lecture. Haber's research on ammonia on the other hand had a direct impact on the war. Without it, Germany might well have been defeated within a few months for lack of explosives. Yet although Haber described in detail the importance of supplies of fixed nitrogen, or "bound nitrogen" as he sometimes called it, for agriculture, he made no reference to the importance of fixed nitrogen for the manufacture of explosives.

The Nobel Prize biography of Haber published at the time of his award was a little more expansive. It noted that Haber's synthesis of ammonia "enabled Germany to prolong the First World War when, in 1914, her supplies of nitrates for making explosives had failed."[21] The biography added that modifications to the process provided ammonium sulfate for use as a soil fertiliser. Haber's contributions to chemical warfare were summarised in one sentence. It stated that after the war started, "he was appointed a consultant to the German War Office and organised gas attacks and defences against them."

Haber carried on working on chemical warfare for the Germans, even after the war, albeit secretly.[22] In 1919, Germany had launched a covert programme to continue the development and production of chemical weapons. The programme was conducted in other countries including the Soviet Union in order to avoid inspections arising from the Versailles Treaty.

Stoltzenberg, who lost an eye in the April 1915 chlorine attack and later helped to construct Germany's mustard gas plants during the war, collaborated with Haber in the programme. In 1922, with Haber's help, Stoltzenberg became involved in a project to build a mustard gas plant for the Spanish army at the *Fábrica Nacional de Productos Químicos* (National Factory of Chemical Products) at La Marañosa in Madrid. Over the next five years, the military used the mustard gas to subdue a rebellion by Moroccan rebels in the Spanish-controlled Rif region of northern Morocco. It is known as the Rif War or Second Moroccan War.

Stoltzenberg's work on the programme ended following the collapse of the joint German–Soviet chemical weapons project in 1926. He went bankrupt. Haber, however, continued with his involvement until 1933, the year Hitler came to power. In the Second World War, Stoltzenberg became a member of the Nazi party. He advocated the use of chemical weapons, but his advice was ignored.[20] No chemical weapons were fired in the war.

A DECISIVE ROLE IN PEST CONTROL

After the First World War, the chemical plants in Germany that had been used for the manufacture of chemical warfare agents were adapted to produce poison gases for pest control. This type of work was not new. During the war, Haber and his gas pioneers had used hydrogen cyanide to tackle insect infestations in flour mills, grain storehouses, railway waggons, hospitals, barracks, and other places. Hydrogen cyanide, also known as hydrocyanic acid or prussic acid, is a highly toxic colourless liquid with an almond-like smell that boils at 26 °C.

Haber's institute had also developed a gas mask to protect pest controllers against the gas during fumigation and in addition designed analytical equipment to detect any gas that remained after the fumigated space had been aired. Haber's expertise in the use of hydrogen cyanide as a fumigant led him to establish the pesticide company Degesch in April 1919. "Degesch" stands for *Deutsche Gesellschaft für Schädlingsbekämpfung* which translates as "German Company for Pest Control."

"Pest control as a scientific discipline and as an industry had been established and institutionalized by war and, in this, one person, Fritz Haber, had played a decisive role," remarked Margit Szöllösi-Janze, professor of contemporary history at the University of Salzburg, Austria, in a review of Haber's work on pesticides published in 2001.[23]

In early 1920, the Dahlem institute developed a pesticide based on methyl cyanoformate, an organic chemical derived from hydrogen cyanide. Known as Zyklon A, the colourless liquid released hydrogen cyanide when it came into contact with water. Haber's group subsequently developed Zyklon B which could be formulated into pellets and was therefore easier and safer to handle. The pellets consisted of a mixture of the hydrogen cyanide, a solid porous material to adsorb the pesticide, a stabiliser, and ethyl bromoacetate, a lachrymatory irritant chemical with a pungent fruity smell that warned people of the presence of the toxic cyanide. The lachrymatory irritant had been used as a tear gas during the war. When the Zyklon B pellets came into contact with air, they released the cyanide as a vapour together with the tear gas with its warning odour.

During World War II, the Nazis employed Zyklon B, without the odorous tear gas, to gas millions of Jews in their extermination camps. Those killed included several members of Haber's family.

OTHER PURSUITS

In 1920, Haber embarked on another project. At the time, the Weimar Republic that had been established in Germany the previous year, was facing considerable problems. Collapse in the value of the German mark had led to hyperinflation and Germany was also required by Article 235 of the Versailles Treaty to make war reparation payments "whether in gold, commodities, ships, securities or otherwise." The payments, equivalent to 20 billion gold marks, were to be paid in instalments between 1919 and 1921.

To help his country out of this desperate economic situation, Haber recruited a team of chemists to investigate the possibility of extracting gold from sea water which they knew contained minute amounts of the precious metal. The project was not a success. After several years of research, the team concluded that the concentration of gold in sea water was far less than originally anticipated and that its extraction was not economically viable. Haber finally abandoned the project in 1926.

In December the following year, after ten years of a marriage that had become increasingly fraught, Charlotte and Fritz divorced. Their nine-year-old daughter Eva and son Ludwig, who had been born in Berlin in July 1921, moved out with their mother into a new home in the centre of Berlin while Fritz remained in his villa in Dahlem. By now, Hermann, Fritz's son by his first wife Clara, had married Margarete ("Marga") Stern (1901–1947) and the couple had moved to the United States.

Over the following years, Fritz Haber was busy carrying out research on fundamental aspects of physical chemistry and at the same time restructuring his institute. He hired top-rate scientists to work there and managed to establish it as a world class centre for pioneering scientific research.

Haber also endeavoured to boost scientific cooperation at both the national and international levels. For example, he played a leading role in gaining Germany's membership of the International Union of Pure and Applied Chemistry (IUPAC). The union had been established immediately after the war by nations that had been "at war with the Central Powers."[24] The founder members were Belgium, France, Italy, the United States, and the United Kingdom. Each member was represented by a "National Adhering Organization," essentially a national federation representing all the societies in the country concerned with pure and applied chemistry.

In 1926, restrictions on membership of the former Central Powers to IUPAC and other international scientific unions were lifted. Before Germany could apply for membership to IUPAC, however, it needed to establish a national federation of its chemical societies and associations. In March 1929, Haber announced that *Verband Deutscher Chemischer Vereine* (Association of German Chemical Societies) had been formed and in December the Germans formally applied for membership of the union. In September the following year, Haber led the German delegation to IUPAC's 10th General Assembly that took place in Liège, Belgium.

It was a poignant location. The German army had invaded Belgium on 4 August 1914 and the next day attacked Liège, a city defended by a ring of 12 forts. It was the opening engagement of the war. The ensuing siege and German artillery bombardments demolished the forts and caused thousands of casualties. The German army eventually captured the city on 16 August and occupied it until the end of the war.

Roger Fennell, author of the union's history, notes: "In view of the damage inflicted on Liège by the Germans during the War, some doubt was expressed as to whether it was a suitable location for the Conference at which a German Delegation would attend for the first time. The Belgian chemists, however, raised no objections."[25]

During the Liège meeting, the IUPAC president, Einar Biilmann (1873–1946), a professor of chemistry at the University of Copenhagen, Denmark, announced that, following a postal vote, Germany had been admitted to the union. IUPAC, which had just five founder members in 1919, now had 30 members. Furthermore, Haber was elected as one of the union's four vice-presidents. He indicated that he "was grateful for the honour afforded to his country but, owing to a purely internal German viewpoint, could not take his place."[26]

IUPAC kept the place open for him. That was in September 1930, the month that Germany held elections and the Nazi Party led by Hitler gained a foothold in the *Reichstag* with a substantial increase in its number of seats. The Nazis became the second largest party in the country. In less than three years, Hitler would become chancellor of Germany and the Nazis the country's largest single party.

SICK, HOMELESS, AND IN EXILE

Haber had always enjoyed "the good life", staying at the best hotels, smoking cigars and eating in the finest restaurants. By the early 1930s, however, he had become portly and lacked energy. He was finding it difficult to sleep, had money worries and his physical health continued to decline. In April, the German government introduced a law requiring civil servants of Jewish descent, with the exception of those who had served in the army in the First World War, to be dismissed. The law covered most

Jewish professors at German universities and the majority of Jewish scientists at the Dahlem institutes. Haber, who had been promoted to captain in the army during the war, was exempt from the edict but he was ordered to dismiss twelve of the scientists in his institute who were Jews.

On 30 April Haber tendered his resignation as director of the institute with a letter in which he wrote: "My tradition requires that when choosing co-workers for a scientific post, I consider only the professional and personal characteristics of the applicant, without regard for their racial makeup."[27] But it was to no avail.

"The remainder of Haber's life is a chronicle of losses," observes his biographer Daniel Charles.[28] He suffered substantial financial losses and at the same time lost his villa, his institute, his faith and his identity as one of the most important and influential people in Germany. His health became increasingly fragile. The fall from the high position he held and the resulting shock proved hard to bear, according to Weizmann who met him several times in London and in Switzerland during the summer of 1933.[29]

Haber allowed five months before his resignation became effective, giving his employers time to recruit a successor and himself time to find other positions for the Jewish staff he was forced to dismiss from the institute. Bosch tried to persuade him to stay and appealed to Hitler on his behalf pointing out that the expulsion of Jewish scientists could seriously damage German science. Hitler angrily replied that Germany could do without chemistry and physics. At the same time, German theoretical physicist Max Planck (1858–1947) raised the issue with Bernhard Rust (1883–1945), the Nazi's Minister of Science, Education and National Culture. Planck was president of the Kaiser Wilhelm Society for the Advancement of Science, the organisation that established the Kaiser Wilhelm institutes. He had also won the 1918 Nobel Prize in Physics for his research on quantum theory. Rust is reported to have replied to Planck's plea that Germany "was done with the Jew Haber."[30]

Albert Einstein (1879–1955), a German-born Jewish scientist and a friend of Haber, had left Germany earlier that year. He was in the United States when Hitler came to power and decided not to return. Before the First World War, Planck and Haber had persuaded Einstein to join them in Berlin and in 1917 Einstein was appointed the first director of the Kaiser Wilhelm Institute for Physics in Dahlem. He won the 1921 Nobel Prize in Physics for his contributions to theoretical physics. In his letters to Haber in 1933, Einstein showed some sympathy for his plight. "I can imagine your inner conflicts," he wrote in May that year.[31]

Over the following months, having found jobs for many of his Jewish colleagues at the institute, Haber travelled to England, France and Switzerland. He received a number of job offers including a standing invitation from Weizmann to go to Palestine and take up a position at a new institute in Rehovot. "He accepted with enthusiasm, and asked only that he be allowed to spend another month or two in a sanatorium," Weizmann noted.[1]

Haber left Germany for good in August 1933 and spent the remainder of his life in an uneasy fluctuating existence in hotel rooms and Swiss sanatoriums in attempts to recover his health. He spent four months living in a hotel in Cambridge following an invitation from Sir William Pope (1870–1939), a chemistry professor at the university (see Chapter 14). Haber was accompanied by his half-sister Else Freyhan who had moved into his villa in Dahlem following his divorce from Charlotte. Else' husband had died several years earlier.

Haber managed to carry out some research in Cambridge but, despite Pope's warm welcome, found it difficult to settle in Cambridge. In January 1934, Fritz and Else left England and travelled by train and boat to Switzerland. By now he was thoroughly tired and exhausted. He died suddenly from a heart attack in Basel on 29 January. His funeral took place at the city's Hörnli cemetery on 1 February. The ashes of his first wife, Clara, were reburied alongside his.

Haber's close friend Willstätter spoke at the ceremony. Willstätter had been appointed chemistry professor at the University of Munich in 1915 but resigned the post in June 1924 in protest against the rising tide of anti-Semitism in Germany. In November 1938 the Gestapo, Nazi Germany's secret police, searched his house and he was later ordered to leave the country. He subsequently settled in Switzerland having lost most of his possessions. He died from a heart attack in 1942.

Shortly after the Second World War other tragedies struck the Haber family. Hermann, the son of Fritz and Clara, and his wife Marga had emigrated to the United States in 1926. Hermann committed suicide in November 1946 and then in June the following year Marga died of Addison's disease. Claire, the eldest of their three daughters, took her own life in 1949 two months before her 21st birthday.

The same year, Fritz and Charlotte's son Lutz, who had settled in England and become a British subject, married Pamela Alice Browne. In 1973 he was appointed Reader in Economic History at the University of Surrey. By then, he had written two books on the history of the chemical industries in Europe and North America.

In 1968, the University of Karlsruhe, formerly the *Technische Hochschule Karlsruhe* where Fritz Haber had taught from 1894–1911, held a ceremony to commemorate the centenary of Haber's birth. During the event, two students climbed onto the stage and displayed a banner with the words: *"Feier für einem Mörder; Haber = Vater des Gaskrieg"* which means "Ceremony for a murderer; Haber = father of gas warfare."

According to Lutz, there was brief silence and then the students disappeared.[32] Some of the audience considered the incident to be an example of student militancy whereas others were annoyed at the disturbance. "On further reflection it seemed to me truly astonishing that after half a century, chemical warfare could still generate so much attention." The incident spurred Lutz to investigate the subject further. When he moved to the University of Surrey he started to work on a book on chemical warfare entitled *The Poisonous Cloud*. It was published in 1986.[14]

Fritz Haber's legacy is extensive. It embraces his nitrogen fixation process which enabled the manufacture of both bread and explosives from air, his work on insecticides, and his efforts to bring back Germany into the international fold of science after the First World War. Yet the Germany that mattered so much to him abandoned him. "It was the tragedy of the German Jew: the tragedy of unrequited love," observed Einstein.[30] His life was a complex mixture of "triumphs, failures and paradoxes," remarks Friedrich. He points out, however, that his love of Germany is no longer unrequited in the country. In 1953, the Kaiser Wilhelm Institute for Physical Chemistry and Electrochemistry in Dahlem that Haber had directed from its opening in 1911 to 1933 was incorporated into the Max Planck Society for the Advancement of Science, an association of some eighty German research institutes, and renamed the Fritz Haber Institute of the Max Planck Society.

REFERENCES

1. C. Weizmann, *Trial and Error: The Autobiography of Chaim Weizmann*, Hamish Hamilton, London, 1949, p. 437.
2. D. Cahan, For science and country, *Nature*, 1995, **375**, 199.
3. C. Flavell-While, Feed the world, *tce (The Chemical Engineer)*, March, 2010, **825**, 54.
4. A. Dronsfield and P. Morris, Who really discovered the Haber process?, *Education in Chemistry*, May, 2007, 85.
5. V. Smil, Detonator of the population explosion, *Nature*, 1999, **400**, 415.
6. B. Friedrich, Gas! Gas! Quick, Boys! How Chemistry Changed the First World War. By Michael Freemantle. (book review), *Angew. Chem. Int. Ed.*, 2013, **52**, 11695.
7. D. Charles, *Between Genius and Genocide: the Tragedy of Fritz Haber, Father of Chemical Warfare*, Pimlico, London, 2006, p. 2.
8. D. Charles, *Between Genius and Genocide: the Tragedy of Fritz Haber, Father of Chemical Warfare*, Pimlico, London, 2006, p. 28.
9. J. E. Coates, The Haber Memorial Lecture, *J. Chem. Soc.*, 1939, 1642.
10. F. Haber, *Thermodynamik technischer Gasreaktionen: Sieben Vorlesungen (Thermodynamics of Technical Gas Reactions: Seven Lectures)*, R. Oldenbourg, Munich and Berlin, 1905.
11. W. Ramsay and S. Young, The decomposition of ammonia by heat, *J. Chem. Soc., Trans.*, 1884, **45**, 88.
12. F. Haber, *Nobel Lectures, Chemistry 1901–1921*, Elsevier, Amsterdam, 1966, p. 326.
13. F. Haber and R. Le Rossignol, Production of ammonia, *US Pat 1 202 995 A*, 1916.

14. L. F. Haber, *The Poisonous Cloud, Chemical Warfare in the First World War*, Clarendon Press, Oxford, 1986, p.146.
15. D. Charles, *Between Genius and Genocide: the Tragedy of Fritz Haber, Father of Chemical Warfare*, Pimlico, London, 2006, p. 175.
16. V. Lefebure, *The Riddle of the Rhine, Chemical Warfare in the First World War*, Collins' Clear-Type Press, London and Glasgow, 1921, p. 31.
17. D. Seyferth, Cadet's fuming arsenical liquid and the cacodyl compounds of Bunsen, *Organometallics*, 2001, **20**, 1488.
18. D. Charles, *Between Genius and Genocide: the Tragedy of Fritz Haber, Father of Chemical Warfare*, Pimlico, London, 2006, p. 155.
19. D. Charles, *Between Genius and Genocide: the Tragedy of Fritz Haber, Father of Chemical Warfare*, Pimlico, London, 2006, p. 157.
20. B. C. Garrett, "Hugo Stoltzenberg and chemical weapons proliferation", *The Monitor* (Center for International Trade and Security at the University of Georgia), 1995, **1**, 11.
21. Fritz Haber – Biographical. *Nobelprize.org*. Nobel Media AB 2013. Web. 26 Feb 2014. http://www.nobelprize.org/nobel_prizes/chemistry/laureates/1918/haber-bio.html.
22. B. Friedrich, Fritz Haber (1868–1934), published in part in *Angew, Chem. Int. Ed.*, 2005, **44**, 3957 and 2006, **45**, 4053.
23. M. Szöllösi-Janze, Pesticides and war: the case of Fritz Haber, *European Review*, 2001, **9**, 97.
24. R. Fennell, *History of IUPAC 1919–1987*, Blackwell Science, Oxford, 1994, p. 11.
25. R. Fennell, *History of IUPAC 1919–1987*, Blackwell Science, Oxford, 1994, p. 31.
26. R. Fennell, *History of IUPAC 1919–1987*, Blackwell Science, Oxford, 1994, p. 49.
27. P. Coffey, *Cathedrals of Science: The Personalities and Rivalries That Made Modern Chemistry*, Oxford University Press, Oxford, 2008, p. 167.
28. D. Charles, *Between Genius and Genocide: the Tragedy of Fritz Haber, Father of Chemical Warfare*, Pimlico, London, 2006, p. 224.

29. C. Weizmann, *Trial and Error: The Autobiography of Chaim Weizmann*, Hamish Hamilton, London, 1949, p. 436.
30. Quoted in Friedrich, ref. 22.
31. D. Charles, *Between Genius and Genocide: the Tragedy of Fritz Haber, Father of Chemical Warfare*, Pimlico, London, 2006, quoted on p. 225.
32. L. F. Haber, *The Poisonous Cloud, Chemical Warfare in the First World War*, Clarendon Press, Oxford, 1986, p. 1.

CHAPTER 13

The World's First Weapons of Mass Destruction

THE BATTLE OF GRAVENSTAFEL RIDGE

Imagine the puzzlement and then the horror of the French and Algerian troops when they saw two clouds of what looked like pale yellowish-green smoke wafting slowly towards them on a gentle breeze from the German lines. They initially suspected that the clouds might be a smokescreen to conceal a German manoeuvre or an infantry attack across No Man's Land, but they could not understand why the clouds were yellowish-green.

As the cloud approached, they began to detect the smell of bleach—just like the smell of the chloride of lime disinfectant powder they scattered around trenches and latrines to kill infectious microorganisms and deter the ubiquitous rats, flies, beetles, slugs and other unpleasant creatures.

The clouds soon rolled over and sank into the trenches engulfing the troops. Some began to choke and gasp for breath. Others dropped to the ground and attempted to bury their faces in the soil to avoid asphyxiation. Those who were able to abandoned their positions in panic to escape the noxious gases. They scrambled out of the trenches only to face the rifle fire and fixed bayonets of German infantry wearing gas masks who had followed behind the clouds of gas.

The Chemists' War, 1914–1918
By Michael Freemantle
© Freemantle, 2015
Published by the Royal Society of Chemistry, www.rsc.org

The gas assault took place late on a sunny spring afternoon on 22 April 1915 at Langemarck, a village some four miles north-east of Ypres, a medieval town in Flanders, Belgium. In the morning that day, the Germans had carried out a heavy artillery bombardment of Allied positions on the Ypres Salient but by early afternoon the front had fallen silent. Then, at 5.00 p.m., the Germans opened the valves of 5730 pressurised metal cylinders containing a total of 168 tons of liquid chlorine.[1] The jets of chlorine released from the cylinder sent clouds of the gas billowing towards the Allied positions. Chemist Fritz Haber, who became known as the Father of Modern Chemical Warfare, directed the operation. The Germans took around five minutes to release the chlorine. Their artillery then followed up with an intense bombardment of the Allied trenches.

The German attack created a four mile breach in the Allied lines around Ypres and their infantry also captured some 2000 prisoners and more than 50 French guns. The use of chlorine gas as a weapon had been an experiment and the Germans were ill-prepared for its success. In particular, they did not have sufficient reserve troops in readiness to exploit the breakthrough. They halted their attack at 8.30 p.m. Had they continued advancing towards Ypres, the Allies would have been put on the back foot and the Germans might well have won the war.

Shortly before midnight, Canadian infantry prepared to counter-attack at an oak plantation in part of the line that had been breached. The plantation was known as Kitcheners' Wood because the French had previously set up field kitchens there. The attack during the early hours of 23 April was successful and the four-mile breach in the line was sealed.

The fighting around Langemarck on 22–23 April 1915 was named the Battle of Gravenstafel Ridge after the village of Gravenstafel and a ridge that formed part of the front line on the Ypres Salient. It was the first engagement in the 2nd Battle of Ypres which continued until 25 May 1915.

At least 800 French, Algerian and Canadian soldiers were killed and over 2000 wounded or gassed in the battle although some reports suggest the number of casualties including fatalities was as high as 5000,[2] or even 6000.[3] Many Germans were also killed or injured by gas as they opened the gas cylinders.

The release of gas at the outset of the battle provided the first example in the history of warfare of the deployment of a weapon of mass destruction. At 4.00 a.m. on 24 April 1915, the weapon was used for the second time when the Germans unleashed 15 tons of chlorine against Canadian infantry holding a position in the Ypres Salient at Saint Julien Wood near Kitcheners' Wood.

By now Canadian medical officers had become aware that the poisonous gas was chlorine. They knew that the gas dissolved in water and therefore suggested to the troops that they protect themselves by soaking cloths, clothing or towels in water and pressing them against the faces. Then another, possibly better, option occurred to them. They remembered that a solution of chlorine in water is weakly acidic and can therefore be neutralised by ammonia, but as ammonia was not available on the battlefield, they surmised that urine could be used in its place. Urine contains urea, a chemical relative of ammonia that also reacts with chlorine. When urine is left to stand for any length of time, the urea reacts with water and breaks down into ammonia and carbon dioxide. The process is catalysed by an enzyme called urease. The Canadian troops therefore, on the advice of their medical officers, famously covered their heads with towels, handkerchiefs, and even socks soaked in their own urine.

The Germans used the weapon five more times during the 2nd Battle of Ypres. Overall, the Germans released almost 500 tons of chlorine from more 20 000 cylinders in cloud-gas operations throughout the battle.

By early May, the Allies had developed other methods of protecting against the gas. Troops were provided with muslin pads stuffed with wads of cotton wool or cotton cloth for use during gas attacks. They dipped the pads into buckets containing solutions of sodium carbonate, sodium bicarbonate, or sodium thiosulfate, before strapping the pads around their faces. The three sodium compounds had several things in common. They were generally available behind the lines. Washing soda is solid sodium carbonate, baking soda is solid sodium bicarbonate, and sodium thiosulfate, or "hypo" as it was called, was used as a photographic fixing agent during the war. All three compounds are highly soluble in water forming alkaline solutions that readily neutralise acidic solutions such as chlorinated water.

The Allies were initially outraged by Germany's cloud-gas operations but at the same time intrigued by their effectiveness. They soon began to develop their own programmes for the production of chemicals weapons and the adoption of more advanced counter-measures to protect against the gases.

Douglas Haig, who at the time commanded the British First Army on the Western Front, was initially sceptical about chemical warfare but subsequently warmed to the idea.[4] The British first employed chlorine gas as a chemical weapon on 25 September 1915, albeit not too successfully, at the Battle of Loos in northern France. In December that year, Haig was appointed Commander-in-Chief of the British Expeditionary Force.

The British and German cloud-gas operations using either chlorine by itself or mixed with other gases continued on the Western Front throughout the war and by the Germans and Russians on the Eastern Front until 1917. In May, June, and July 1918, for example, the British launched several cloud-gas operations against the Germans between La Bassée and Loos, a sector of the Western Front near Lens in northern France.

THE CHEMICAL WEAPONS' RACE BEGINS

Chlorine was not the first toxic gas to be used in the First World War. The French used grenades filled with ethyl bromoacetate, a tear gas, in August 1914 and Germans fired tear gas shells known as T-shells at Russian troops on the Eastern Front during the Battle of Bolimów, near Warsaw, on 31 January 1915. T-shells contained "T-Stoff" a lachrymatory mixture of xylyl bromide and benzyl bromide, but these gases are not lethal and have therefore never been considered as weapons of mass destruction.

Chlorine was different, however. The choking gas attacks and irritates the respiratory tract causing a variety of symptoms and if inhaled in high concentrations, it kills. After the initial chlorine gas attack on 22 April 1915, the Allies rapidly developed improved head coverings that provided protection against the gas. The "British Smoke Hood," as it was called, began to be distributed to troops in France in May 1915 and by early July 1915 all British troops on the Western Front had been issued with them. French troops also used them. They were made of a flannel bag soaked in a solution of glycerine, sodium carbonate

Figure 13.1   British smoke hood (author's own, 2012, Imperial War Museums, London.)

and sodium thiosulfate. The glycerine prolonged the life of the solution. The bag incorporated a rectangular eyepiece (Figure 13.1).

Soon after the Allies had developed methods of protecting against chlorine, the Germans introduced phosgene, an even more deadly choking agent that could penetrate the hoods. They called it "D-Stoff" and first used it in a cloud-gas operation against the British in Flanders on 19 December 1915. Phosgene is denser than chlorine and consequently not readily carried by the wind. The gas was therefore mixed with chlorine to form a cloud that was light enough to be blown towards Allied trenches. The Allies also used the mixture, which they called "White Star" gas, in cloud-gas attacks.

As the war progressed, the Central Powers and Allies engaged in a leap-frogging contest of gas offense and gas defence. As soon as a novel chemical warfare agent was introduced by one side, the agent was identified by the other side, and counter measures rapidly developed to protect against it. In January 1916, for example, the British introduced the PH (phenate-hexamine) helmet that offered protection not only against phosgene, but also against chlorine and the tear gas that continued to be deployed on the Western Front. The flannel used to make the helmet was soaked in a solution of phenol, hexamine, caustic soda, and glycerine.

The success of cloud-gas operations depended on wind conditions. If the wind was strong enough and in the right direction, the gas released from cylinders drifted towards enemy territory dispersing throughout the atmosphere over enemy lines and beyond. As chlorine is over twice as dense as air, the clouds clung to the ground as they rolled forward sinking into trenches, penetrating dugouts, and seeking out spaces that bullets, shrapnel balls, and red-hot sharp shell fragments could not reach. However, because the clouds of gas were so pervasive, they lacked precision. The clouds gassed and killed indiscriminately, poisoning not only enemy soldiers but also innocent civilians. The chlorine gas cloud that passed over and into the trenches at Langemarck, for example, continued towards Ypres and engulfed civilians who had turned and fled in panic towards the town along with the fleeing soldiers.

Artillery shells filled with poisonous chemicals, on the other hand, could be aimed precisely at specific physical targets in the enemy's lines. However, liquid chlorine by itself was too volatile to be loaded into shells or other projectiles such as trench mortar bombs or grenades.

The first lethal chemical to be used in gas shells was phosgene (Figure 13.2). The French introduced the shells in February 1916 at the start of the Battle of Verdun. Phosgene has a low boiling point of just 8 °C which means that it is a gas at room temperature under normal conditions. It therefore had to be kept refrigerated as a liquid below this temperature until it was poured into shells, cylinders, or bombs.

Diphosgene, an organic chemical and lethal lung irritant with a similar level of toxicity to phosgene, boils at 128 °C. At room temperature it is an oily liquid that can be readily poured into projectiles. The Germans first used the chemical in

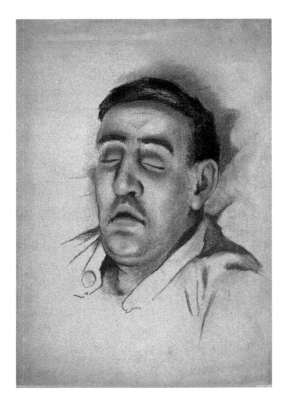

Figure 13.2   Face of soldier suffering from the effects of phosgene gas poisoning (Wellcome Library, London).

artillery shells in May 1916 against the French at the Battle of Verdun.

A couple of months later, the French introduced another potentially lethal type of chemical warfare agent. Their so-called Vincennite shells contained hydrogen cyanide, a relatively simple chemical that, once it gets into the blood stream, prevents the body's cells from using oxygen. They began firing the shells at the Battle of the Somme on 1 July 1916, the first day of the battle. Vincennite shells were not successful, however. Hydrogen cyanide is highly volatile and so it was impossible to generate lethal concentrations of the gas when the shells exploded.

Chloropicrin was the next deadly chemical warfare agent to appear on the battlefield. The chemical is less toxic than phosgene but more toxic than chlorine. Although used primarily as a lung irritant, it was also a strong tear gas and caused nausea, vomiting, and diarrhoea. It could penetrate the gas masks used to protect against chlorine and phosgene and was relatively easy to manufacture. The Russians first deployed the chemical in August 1916.[5] The Allies and Germans subsequently used the gas either by itself or with other chemicals in cylinder cloud-gas operations, artillery shells, and trench-mortar bombs.

AN UNUSUAL FAILURE

In July 1917, the Germans launched a completely different class of chemical warfare agent, the sternutator. Such agents were employed specifically to irritate the nasal passages and rapidly induce sneezing, nausea, vomiting and headaches. The first chemical to be used for this purpose was diphenychlorarsine. The arsenic-containing solid was packed into shells with a high explosive. When the shells exploded, the solid was dispersed as a dust.

But it was not lethal in the concentrations achieved on the battlefield, and nor was it intended to be. So why introduce this non-lethal agent when gas shells filled with lethal chemicals like diphosgene were available?

The answer lies with the gas masks in use at the time. With the rapid development of chemical warfare programmes in Britain, France, and Russia, gas masks that provided good protection against choking gases such as chlorine, phosgene and other toxic chemicals were being produced in their hundreds of thousands.

The German sternutator shells were designed to eject fine dust particles when the shells exploded. The particles would then quickly penetrate gas masks and induce the Allied troops to sneeze and vomit. The troops would consequently be forced to remove their masks and expose themselves to lethal chemicals released from other types of gas shells that were being fired simultaneously with the sternutator shells.

That was the theory. In practice the shells didn't work. The dust particles produced when the shells burst were too big to penetrate the Allies' gas masks. Furthermore, the shells were not fused to explode and release the dust where it could be inhaled, that is just above the ground where the troops were standing. The failure to find a way of dispersing diphenylchlorarsine "stands out as one of the few technical mistakes in chemical warfare that the Germans made" during the First World War according to Augustin Prentiss (1915–2009), a lieutenant colonel in the Chemical Warfare Service of the United States Army.[6]

THE NASTIEST OF THEM ALL

In July 1917, the same month that they launched their sternutator shells, the Germans also introduced another chemical warfare agent: dichloroethyl sulfide. Commonly known as mustard gas or sulfur mustard, it was the most notorious all the poisons used in the war. The gas had one distinct advantage over all the other chemical warfare agents that had been used in the Great War to date. It attacked the whole body and not just the eyes and lungs. Gas masks that had been designed and manufactured in their millions to protect against tear gases and choking agents offered only limited protection against mustard gas. The poisonous chemical effectively side-stepped gas masks.

Mustard gas is principally a vesicant or blister agent. Its role as a lung-injurant is subsidiary. The chemical is actually not a gas but an oily liquid under normal battlefield conditions. It is yellow or brown depending on its purity and gives off a sweet agreeable odour reminiscent of garlic or the mustard plant. The smell is sometimes so weak, however, that it is barely detectable by the human nose.

Soldiers who came into contact with the liquid carried it around on their clothing, shoes, and boots. The vapours released by the liquid, particularly in warm rooms, dugouts, and trenches, penetrated army uniforms, including rubber and leather boots. Troops often did not realise they had been exposed to the chemical as they could not smell it in the stench of the battlefield or trenches. It had a delayed effect and so victims did not experience the symptoms of exposure until several hours later.

Mustard gas attacks and burns skin forming small fluid-filled blisters, known as vesicles, that slowly grow into much larger blisters. The gas destroys body tissue and damages eyes causing blindness. If inhaled it leads to internal bleeding. Those exposed to high levels of mustard gas sometimes suffered agonising deaths from their horrendous injuries over a period of four or five weeks. "The pain was so bad that most soldiers had to be strapped to their beds," according to science historian Patrick Coffey.[7]

Unlike chlorine and phosgene which are rapidly dispersed in the atmosphere, mustard gas is highly persistent. When a mustard gas shell exploded, the liquid immediately formed an aerosol, a suspension of liquid droplets in air. The aerosol than sank to the ground or the bottoms of trenches and in some cases could lie around for days on end or even weeks in cold weather. Because of its persistency, mustard gas proved particularly effective in defensive operations and in retreat as it deterred enemy troops from attacking across ground contaminated with the oily liquid.

The Germans began firing mustard gas shells at the British front lines near Ypres on the night of 12/13 July 1917. Within ten days they had fired more than one million shells containing a total of about 2500 tons of the vesicant.[8] The attacks resulted in a heavy toll of casualties

Haber was impressed. In 1916, he had looked into the possibility of using mustard gas as a chemical weapon but dismissed the idea as he believed its toxicity was too low. At the time, the blistering properties of the chemical were well known but not considered to be significant in chemical warfare. After the Ypres mustard gas attacks in July 1917, Haber revised his opinion claiming mustard gas to be a "a fabulous success." The agent later became known as the "king of the battle gases."[9]

Within a short time of these attacks, Allied chemists identified the chemical and started to develop their own industrial processes for its manufacture. The French were quickest off the mark and began firing mustard gas shells on the Western Front on 16 June 1918. The British fired them for the first time at the end of September on the opening day of the Battle of St Quentin Canal in France. It was one of the many battles that comprised the Allies' Hundred Days Offensive that led to the armistice between the Allies and Germany on 11 November 1918.

Another vesicant developed very late in the war was even more deadly than mustard gas. Chlorvinyldichlorarsine, named Lewisite after the American chemist Winford Lee Lewis (1878–1943) who worked on its development, contained arsenic and was highly toxic. Like mustard gas, Lewisite is an oily liquid that blisters the skin and injures the lungs. It had two distinct advantages over mustard gas. First, it was much more volatile and therefore non-persistent on the ground and in trenches.

Second, it was much faster acting. For these reasons, it could be used for offensive operations. On the downside, gas masks protected against it and furthermore it did not penetrate army uniforms as well as mustard gas.

Even so, the United States manufactured the agent in secret with the idea of using military aircraft to spray the chemical over cities in Germany. In anticipation of this deadly precipitation, Lewisite was dubbed "the dew of death." The Americans shipped a consignment of 150 tons of the deadly oily liquid to Europe in November 1918 but they didn't have a chance to use it. The armistice between the Allies and Germany was signed before the ship arrived and the consignment was "unceremoniously dumped into the ocean."[10]

MASS PRODUCTION, MASS SLAUGHTER

In his treatise on chemical warfare, published in 1937, Prentiss provides a breakdown of the battle gases, as he called them, produced and used in the First World War.[11] Forty-six different gases were employed. In terms of quantities produced, the major players were Germany, France, and Britain who manufactured approximately 68 000 tons, 37 000 tons, and 26 000 tons respectively. The United States, Austria, Italy, Russia were relatively minor players each producing between 3500 tons and 6500 tons. In all, some 150 000 tons of battles gases were manufactured by the belligerent nations. Choking gases accounted for more than 123 000 tons, vesicants just over 13 000 tons, sternutators over 7000 tons, and tear gases 6000 tons. A total of 125 000 tons of poison gases were used in the war, the remaining 25 000 tons being "left on hand unused," according to Prentiss.

Prentiss details more than 70 major gas attacks carried out by the Germans against the British, French, and Americans throughout the war.[1] Britain's first use of gas at the Battle of Loos on 25 September 1915 encountered a number of problems. The artillery did not have correctly-fitting spanners to open all the chlorine cylinders, some of the pipes used to dispense the gas leaked, and the wind was not always strong enough to carry the gas. The British use of poison gas "never had such prominence again," according to British military historian Gary Sheffield.[12]

Prentiss lists just 17 major British attacks between September 1915 and July 1918. British military historian Gordon Corrigan, however, indicates that the British launched 768 gas attacks over the same period using cylinders, projector drums, gas bombs, and shells.[13] He notes that the British used 5700 tons of gas of various types, an amount notably less than that given by Prentiss. What is not in doubt is that Britain used poison gas extensively during the war.

Gas offence and gas defence developed hand in hand throughout the war. In the three and half years of the war from May 1915 to November 1918, the British government issued some 50 million gas masks of various types to protect its army in Belgium and France against toxic gases.[14]

By the time of the 3rd Battle of Ypres, which was fought during the second half of 1917, gas masks had become relatively sophisticated compared with the primitive ones employed during the 2nd Battle of Ypres in 1915. The later masks were designed to protect against a variety of poison gases. They typically incorporated a canister filled with granules of soda lime, activated charcoal, and sodium permanganate or potassium permanganate (Figure 13.3). Soda lime, a mixture of calcium hydroxide and a small amount of sodium hydroxide, is an alkaline material that neutralises acidic gases such as chlorine. Activated charcoal is a sponge-like form of charcoal with numerous pores and consequently has a vast internal surface area. Some, but not all, the poison gases become trapped inside the charcoal's pores. The permanganate is a strong oxidant that destroys a variety of toxic gases.

Despite the extensive use of gas masks and other forms of protection, hundreds of thousands of troops were gassed during the war. The severity of the gassing depended on the nature and toxicity of the gas, its concentration, and the volume that had been inhaled. Gas casualties were evacuated from the frontline as rapidly as possible. In less severe cases, casualties made their own way to first aid posts, dressing stations, and casualty clearing stations. Stretcher bearers carried others.

Inhalation of choking gases inflamed the lungs and other parts of the respiratory tract resulting in symptoms such as coughing, nausea and vomiting. Fluids secreted in the tract accumulated in the lungs, deprived the heart of oxygen, and were likely to drown the victim.

Figure 13.3   German gas mask (Science Museum, London, Wellcome Images).

Treatment varied. Some troops recovered quickly as soon as they could breathe in fresh air. For less critical cases, emetics proved useful, reported British doctor Leonard Hill in a paper on gas poisoning read before the Medical Society of London and published in December 1915.[15] "Half a pint of salt and water or 8 grains of copper sulphate, followed by large draughts of lukewarm water, are recommended," he observed. "Warmth and good nursing might pull a man through." Atropine or other drugs were sometimes used to relieve symptoms and minimise the release of fluids.

But if the victim did not recover within a few hours, there was a good chance he would die from heart failure resulting from oxygen deprivation within a day or so. The Germans had oxygen cylinders available to treat gas casualties as early as the summer of 1915,[16] and the British soon recognised the value of oxygen therapy. "In the treatment of the chlorine and phosgene cases by far the most important thing was the supply of oxygen," observed Major-General Sir Wilmot Herringham (1855–1936), a British doctor in the Royal Army Medical Corp (RAMC) who pioneered the study and treatment of gas poisoning in the war.[17]

He noted that death rarely occurred if the patient survived the third day.

Another RAMC doctor, Major Walter Broadbent (1868–1951), observed that men suffering from gas poisoning who had been sent back to Britain with breathing difficulties did not like the use of oxygen.[18] They preferred an inhalation of tincture of benzoin from a steam kettle or even better a linseed poultice covering the entire back. Benzoin in this context should not be confused with the synthetic organic chemical of the same name. Tincture of benzoin is a solution of benzoin resin in alcohol. The resin is extracted from the bark of certain species of trees.

Troops exposed to mustard gas were quickly transferred from the frontline to casualty clearing stations where they removed their clothing and washed themselves down with warm water and soap. Saline baths were also used.[19] Victims were then treated for burns and inflammation of the throat and lungs. Mustard gas resulted in some 125 000 British casualties between July 1917, when the gas was first used, and the end of the war. Less than 2000 of the cases proved fatal.

Overall, chemical attacks of one sort or another caused just 3% of the almost three million British casualties in the war. Artillery shelling and rifle fire accounted for 58% and 39%, respectively. Chemical warfare led to some 1.5 million casualties among all the belligerent nations of which less than 1% proved fatal. Even so, that is 150 000 deaths caused by gas.

OUTRAGE AND SUPPORT

The success of the German gas attack on 22 April 1915 surprised both the Germans and the Allies and caused panic and outrage among the Allied nations. An international peace treaty signed by Britain, France, Germany and other nations at a peace conference held at The Hague in the Netherlands in 1899 and the follow-up Hague Convention of 1907 had effectively banned chemical weapons. The signatories agreed "to renounce the use of all projectiles, the sole object of which is the diffusion of asphyxiating or deleterious gases." It should be noted, however, that gas shells had yet to be developed in 1899 and that cylinders rather than projectiles were used to disperse the toxic chlorine gas in the attack on 22 April.

The opposing armies deployed chemical weapons during the First World War not just to kill enemy troops with poison, but also to harass, confuse, disorient and terrorise the soldiers in the hope that they would be driven from their trenches. Furthermore, the use of gas was also intended to incapacitate troops who remained in the trenches. Gas masks slowed down their ability to fight and communicate with one another. Finally, care of victims suffering from the effects of toxic gases tied up valuable medical resources.

But whether chemical warfare contributed to any great extent to the final outcome of the Great War is questionable. Ludwig Haber, son of Fritz Haber, considered gas a failure. In his book *The Poisonous Cloud: Chemical Warfare in the First World War*, he observes that "poison gas was an ingenious attempt to overcome trench warfare."[20] But although it was not expensive to produce and capable of further development, it was largely a waste of effort, he concluded. In the final year of the war, "this unreliable and ineffective weapon" helped the Germans more than the Allies, "but it did not win them a battle, let alone give them victory."

Although widespread public indignation and revilement at the ghastly, frightening, and insidious nature of chemical weapons continued throughout the war and afterwards, there were many influential figures on both sides, especially in science, politics, and the military, who encouraged their use whole heartedly. For instance, after initial doubts, Haig supported the development and use of chemical weapons. A few weeks before the Battle of Loos in September 1915, he had been impressed by a demonstration of the discharge of a cloud of chlorine gas from cylinders, Ludwig Haber remarked.[21]

French marshal Ferdinand Foch (1851–1929), who was Commander-in-Chief of the Allied Armies at the end of the war, also approved of their use. Following the war, he observed that the ability of aircraft to carry increasing weight furnished a new method for the large-scale military use of poison gas. "Chemical warfare thus acquires the power to produce more terrible effects over much larger areas [and] must necessarily enter into our forecast of the future, and our preparations for it." He expressed these thoughts in the introduction to *The Riddle of the Rhine: Chemical Strategy in Peace and War*, a book about the development of chemical warfare.[22] The author of the book was British chemist Victor Lefebure who commanded two Special Companies of the Royal Engineers when they carried out a successful cylinder attack with White Star gas against the Germans at Nieuport, a town in Flanders, on the night of 5–6 October 1916.[23] The companies were established by the British in May 1915 to conduct chemical warfare operations and subsequently combined and expanded into the Special Brigade of the Royal Engineers (see Chapter 2).

In 1917, Lefebure, a fluent French speaker, was appointed, with the rank of major, as the British Chemical Warfare Liaison Officer with the French. His book on chemical warfare was published in 1921 and aimed to persuade the League of Nations, the precursor to the United Nations, not to ban chemical weapons. He was unsuccessful, however. An international protocol prohibiting the use of both chemical and biological weapons was signed in Geneva in 1925 and came into force in 1928. It is known as the Geneva Protocol.

Lefebure's book also referred to a report by British chemist Harold Hartley, another proponent of the use of chemical weapons. Hartley was a lecturer in chemistry at Oxford University, served in the British Army, and by the end of the war had been promoted to the rank of brigadier general and was Controller of the Chemical Warfare Department at the British Ministry of Munitions. He observed that the battlefield death toll from poison gas was much lower than from other causes and a much smaller proportion of those injured by gas suffered any permanent disability.[24] "There is no comparison between the permanent damage caused by gas, and the suffering caused to those who were maimed and blinded by shell and rifle fire," he commented in his report to the British Association, now known as the British Science Association.

Winston Churchill, who from July 1917 to January 1919 was Minister of Munitions in the British coalition government led by David Lloyd George, was another keen advocate of the use of chemical weapons. In May 1919 he claimed that he did not understand "this squeamishness about the use of gas."[25] He pointed out that it was not only necessary to use the most deadly gases. Gases could also be deployed to inconvenience the enemy, spread terror and yet not leave serious permanent effects on

Figure 13.4 From left: Sir Harold Hartley, English botanist Sir Edward Salisbury (1886–1978), and Nevil Vincent Sidgwick (RSC Library).

most of those affected. He remained convinced of their usefulness right through to the Second World War. In July 1944, when he was Secretary of State for Defence, he wrote a memorandum to the War Office requesting that Germany be drenched with poison gas.

Across the Atlantic, Amos Fries (1898–1963), who was appointed Chief of the United States Army's Chemical Warfare Service (CWS) in 1920, took a similar view of chemical warfare. "Considering its power it has no equal," he noted in an article on the topic that appeared in the May 1920 issue of the *Journal of Industrial and Engineering Chemistry*, a monthly journal published by the American Chemical Society (ACS).[26]

The ACS followed up a few months later with a promise to support the work of the CWS. "Upon invitation of General Fries, the American Chemical Society has pledged the active aid of its 15 000 civilian members in the successful development and prosecution of the work of the Chemical Warfare Service," the journal's editor Charles H. Herty (1867–1938) announced in his opening address at the 6th National Exposition of Chemical Industries held in New York, September 1920.[27] Herty was ACS president in 1915 and 1916.

After listing 15 members of an ACS committee that would provide a connecting link with the CWS, Herty added: "Chemical warfare has come to stay. The effectiveness of gas in warfare has been proved by the fact that one-third of the total hospital cases in our Army was due to gas; its inhumanity has not proved itself in the light of history, for of this third our medical records show that the very great majority completely recovered, a far greater proportion than of those who were wounded by shot and shell."

Finally, how did the "Father of Modern Chemical Warfare" feel after the First World War about the use of chemical weapons having personally witnessed the death, injury and suffering they caused? In an address to German officers in 1920 he pointed out that these weapons were intended to have not just a physiological impact on the enemy but also a psychological impact.[28] They owed their ultimate success to their ability to overpower the enemy's moral. "Chemical warfare is certainly no more horrible than flying pieces of steel; on the other hand, the percentage of mortality from gas injuries is smaller," Haber said. His views were similar to those of Haig, Foch, Hartley, Churchill, Fries and Herty.

Such comments by eminent British, French, American and German figures during and after the war endorsing the use of chemical weapons may seem astounding to us nowadays, especially when we remember the horrors of Halabja. On 16 March 1988, Saddam Hussein's regime in Iraq dropped bombs containing various poisonous chemicals, including mustard gas, on the Kurdish town killing some 3000 to 5000 people. Even more people were wounded.

More recently, people all around the world were shocked and appalled by the chemical attack on civilians in the Ghouta region near Damascus in the Syrian civil war. Various sources reported that between 350 and 1500 people died on 21 August 2013 when rockets containing sarin struck the area. Sarin is a highly volatile organophosphate nerve agent that inhibits breathing and results in death by asphyxia. News images and footage showed numerous dead bodies and pallid and terrified men, women, and children struggling for breath.

However, by focusing on the horrors of chemical weapons there is a danger that we overlook the dreadful consequences of the use of conventional weapons. The sinking of the British ocean liner RMS (Royal Mail Ship) *Lusitania* provides just one example of the latter. On 7 May 1915, just about two weeks after the first chlorine gas attack on the Ypres salient, a torpedo fired by a German U-boat hit the liner. It sank claiming the lives of 1198 passengers and crew. A single torpedo fired by a submarine might therefore also be regarded as a weapon of mass destruction.

REFERENCES

1. A. M. Prentiss, *Chemicals in War: A Treatise on Chemical Warfare*, McGraw-Hill, New York, 1937, p. 663.
2. E. Croddy with C. Perez-Armendariz and J. Hart, *Chemical and Biological Warfare: A Comprehensive Survey for the Concerned Citizen*, Copernicus Books, New York, 2002, p. 144.
3. J. Lee, *The Gas Attacks Ypres 1915*, Pen & Sword, Barnsley, 2009, p. 21.
4. G. Sheffield, *The Chief: Douglas Haig and the British Army*, Aurum Press, London, 2011, p. 125.
5. A. M. Prentiss, *Chemicals in War: A Treatise on Chemical Warfare*, McGraw-Hill, New York, 1937, p. 161.
6. A. M. Prentiss, *Chemicals in War: A Treatise on Chemical Warfare*, McGraw-Hill, New York, 1937, p. 206.
7. P. Coffey, *Cathedrals of Science: The Personalities and Rivalries That Made Modern Chemistry*, Oxford University Press, Oxford, 2008, p. 109.
8. R. Evans, *Gassed*, House of Stratus, London, 2000, p. 40.
9. D. Charles, *Between Genius and Genocide: the Tragedy of Fritz Haber, Father of Chemical Warfare*, Pimlico, London, 2006, p. 68.
10. E. Croddy with C. Perez-Armendariz and J. Hart, *Chemical and Biological Warfare: A Comprehensive Survey for the Concerned Citizen*, Copernicus Books, New York, 2002, p. 102.
11. A. M. Prentiss, *Chemicals in War: A Treatise on Chemical Warfare*, McGraw-Hill, New York, 1937, p. 661.
12. G. Sheffield, *The Chief: Douglas Haig and the British Army*, Aurum Press, London, 2011, p. 372.
13. G. Corrigan, *Mud, Blood and Poppycock*, Cassell, London, 2003, p. 173.
14. A. M. Prentiss, *Chemicals in War: A Treatise on Chemical Warfare*, McGraw-Hill, New York, 1937, p. 534.
15. L. Hill, Gas Poisoning, *Br. Med. J.*, 1915, **2**, 801.

16. G. Hartcup, *The War of Invention: Scientific Developments, 1914–18*, Brassey's Defence Publishers, London, 1988, p. 174.
17. W. Herringham, Medicine in the War: A Retrospective Sketch, *Br. Med. J.*, 1919, **1**, 20.
18. Y. McEwen, *"It's a Long Way to Tipperary" - British and Irish Nurses in the Great War*, Cualann Press, Dunfermline, 2006, p. 97.
19. K. Brown, *Fighting Fit: Health, Medicine and War in the Twentieth Century*, The History Press, Stroud, 2008, p. 119.
20. L. F. Haber, *The Poisonous Cloud, Chemical Warfare in the First World War*, Clarendon Press, Oxford, 1986, p. 279.
21. L. F. Haber, *The Poisonous Cloud, Chemical Warfare in the First World War*, Clarendon Press, Oxford, 1986, p. 55.
22. V. Lefebure, *The Riddle of the Rhine, Chemical Warfare in the First World War*, Collins' Clear-Type Press, London and Glasgow, 1921, p. 9.
23. J. Davidson Pratt and S. Smiles, Obituary: Victor Lefebure, *J. Chem. Soc.*, 1948, 394.
24. V. Lefebure, *The Riddle of the Rhine, Chemical Warfare in the First World War*, Collins' Clear-Type Press, London and Glasgow, 1921, p. 236.
25. B. Macintyre, Shrapnel kills just as surely as a gas bomb, *The Times*, September 6, 1913, p. 33.
26. A. A. Fries, Chemical warfare, *Ind. Eng. Chem.*, 1920, **12**, 423.
27. C. H. Herty, Opening address, *Ind. Eng. Chem.*, 1920, **12**, 956.
28. F. Haber, Chemistry in War, *J. Chem. Educ.*, 1945, **22**, 526.

Pope and the Mustard Agents

A NEW OFFENSIVE MATERIAL

In July, 1917, the Germans used a "new offensive material" against the Allies, and "with very great success," noted Sir William Jackson Pope (1870–1939) in his presidential address delivered at the annual general meeting of Britain's Chemical Society on 27 March 1919.[1] Pope was president of the society from 1917 and 1919 and had played a leading role in British efforts to develop processes for the manufacture of chemical weapons for the Allies on the Western Front. Soon after the Germans started to bombard Allied lines with shells filled with the new offensive material, British and French chemists identified the chemical and before long Britain and France embarked on their own programmes to manufacture the material.

Along with many other chemists at the time, Pope considered the use of this new toxic agent to be humane because it was designed to incapacitate large numbers of enemy troops rather than kill them. "This substance, the so-called 'mustard gas,' has but little odour, and exposure to it causes comparatively few fatalities; inhalation of, or contact with, its vapour gives rise to acute pneumonia, to the production of painful sores, and to temporary or even permanent blindness," Pope went on to say in his address. "Whilst, as has been stated, the actual mortality is

The Chemists' War, 1914–1918
By Michael Freemantle
© Freemantle, 2015
Published by the Royal Society of Chemistry, www.rsc.org

low, and the use of the substance may to this extent be described as humane, the casualties produced are very numerous; slight exposure to a material so toxic and so difficult to detect leads, in general, to six weeks in hospital."

Pope, yet another keen advocate of the use of chemical weapons, was one of the most distinguished British chemists in the early decades of the 20th century. He was born and educated in London and began to take an interest in chemistry as a young boy. By the time he was 12 years old, he had acquired a collection of chemicals and simple apparatus which he kept in his bedroom cupboard. He also had a talent for learning foreign languages. "He became proficient in both French and German whilst still in his teens and at the age of 15 was using a book of German mathematical tables, which was lying on his study desk at the time of his death," notes fellow British chemist Gerald Moody (1864–1943) in his obituary of Pope.[2]

Figure 14.1   William Jackson Pope (Royal Society of Chemistry Library).

Moody and Pope were both early students of Henry Edward Armstrong (1848–1937), a chemistry professor at a college that was to become Imperial College London. Armstrong was an inspirational teacher and was one of the most outstanding figures in the history of British chemistry. He carried out pioneering work on the chemistry of benzene, naphthalene and their derivatives.

In 1897, Pope was appointed head of the chemistry department at the Goldsmiths' Institute in London and four years later he moved to Manchester Municipal School of Technology to become professor of chemistry and head of the chemistry department. He was elected professor of chemistry at Cambridge University in 1908, a position he held until his death in 1939.

FRUSTRATED BY INCOMPETENCE AND INEFFICIENCY

Pope's research interests included crystallography and the stereochemistry of organic compounds. During the First World War, he served as a member of the panel of consultants of the Board of Invention and Research set up by the British government in July 1915 to assess invention proposals put forward by the public. "He devoted himself wholeheartedly and with consummate skill to those chemical problems arising from the war, towards the solution of which his vast knowledge, acquired by wide reading, proved of utmost value," observes Moody.

Even so, Pope was highly critical of and outspoken about attitudes to science among the higher echelons of the British science policy establishment during the war.[3] He complained of ignorant authorities and the contempt for science by influential circles of the British establishment. He blamed the horrors experienced in the early days of the war on the way experts were made to be subservient to administrators. Furthermore, he considered British arrangements for chemical warfare to be highly inefficient.

Following the German chlorine gas attack in April 1915, the British rapidly began to develop their own chemical warfare programme, but its organisation proved hierarchical, complex and somewhat chaotic in nature. In late 1916, for example, the organisation of the programme within the Ministry of Munitions consisted of numerous components including separate departments for explosives, munitions, trench warfare, and trench warfare research.[4] The trench warfare research department had two divisions, one of which, the research division, had a chemical advisory committee with four sub-committees. The advisory committee advised the trench warfare supplies committee which was part of the operation division of the chemical section, one of ten sections of the trench warfare department.

And it didn't stop there. The War Office had separate responsibility for anti-gas measures. Its medical department had an anti-gas department with an anti-gas supply committee which in turn had a committee that dealt with supplies of components, inspection, and research and development.

During 1917, efforts were made to reorganise the structure and "end the absurdities" of British chemical warfare organisation, as Ludwig ("Lutz") Haber put it in his book on chemical warfare in the First World War.[5] The author quotes a letter Pope wrote to British chemist Harold Hartley on 20 August that year. Hartley, a lecturer in physical chemistry at Oxford University and at the time a lieutenant-colonel in the British Army, was serving as assistant director of gas services at the army's general headquarters in France. Pope protested to Hartley that even though Britain was full of excellent chemists, as good as any in Germany, chemists outside London were not being allowed to participate in work on gas warfare. Chemical warfare research in Britain was "being run in the spirit of the little village publican doing his two barrels a week," he wrote.[6]

In September that year, the chemical advisory committee established an advisory panel of twelve scientists, but Pope was not included even though he and other staff in the chemistry department at Cambridge were making substantial contributions to the British chemical war effort. Towards the end of the year the Ministry of Munitions established the Chemical Warfare Department but the situation barely improved. "The time wasted by mismanagement and the toleration of incompetence could not be made up," according to Haber.[7]

Pope's frustration at the way chemical warfare was organised in Britain did not dampen his efforts in the laboratory to contribute to the British war machine, however. Together with his university colleagues, he investigated new sources of aromatic hydrocarbons such as toluene that were needed for the manufacture of high explosives like TNT and amatol (a mixture of 40% ammonium nitrate and 60% TNT).

In another project, Pope worked with William Hobson Mills (1873–1959) on the development of new synthetic dyes known as cyanines. These nitrogen-containing organic compounds were sensitive to light and potentially useful for improving the sensitivity of the photographic plates used in aerial reconnaissance. Mills was a lecturer in organic chemistry at Cambridge. He should not to be confused with William Mills (1856–1932), the British inventor of the grenade known as the "Mills bomb," tens of millions of which were produced and used by the Allies in the First World War.

Pope also worked on the preparation of phosgene. The Germans first used the poison gas against the British in Belgium in December 1915. The French employed the gas at Verdun in February 1916 and the British at the Battle of the Somme later that year.

In other work, Pope and colleagues carried out research on the synthesis of arsenic-containing organic compounds known as arsenicals. Such compounds were introduced by the Germans in the summer of 1917. They were employed as sternutators, which are chemical warfare agents that induce sneezing and vomiting (see Chapter 13).

Of all his chemical warfare activities, Pope is most famous for his work on the development of an efficient process for the production of mustard gas, otherwise known as sulfur mustard or dichloroethyl sulfide. Although the process for making the chemical was of considerable interest to the scientific community when Pope carried out the work, it was not until well after the end of the war that the details of the process were published.

SMELL OF MUSTARD OIL, TASTE OF HORSE-RADISH

Like all the chemical warfare agents widely used in the First World War, mustard gas was first synthesised many years before the war. Belgian-born French chemist César-Mansuète Despretz (1798–1863) is reported to have been the first person to prepare the toxic chemical. In 1822, he described some of the characteristics of the product of the reaction between ethene and sulfur dichloride—one of two compounds consisting of sulfur and chlorine, the other being sulfur monochloride.[8] Despretz observed that the product was oily and had a bad smell but he did not mention its irritating effects.[9]

English chemist Frederick Guthrie (1833–1886) carried out similar experiments many years later and synthesized a compound which he found to be extremely poisonous. It was mustard gas, although it was not called that until the First World War.

Guthrie, was born in London and studied chemistry at University College London before travelling to Germany where he continued his studies and obtained a PhD. He then returned to Britain and at the time of his research on mustard gas and related compounds was published, he was working at Edinburgh University as an assistant to Lyon Playfair (1818–1898), an eminent Scottish chemistry professor. In 1853, Playfair had suggested that shells filled with cyanide or phosphorus should be fired at the Russians in the Crimean War (1853–1856) in which Russia fought against Britain, France, Piedmont (a region of Italy), and Turkey. The British Admiralty and British War Office turned down Playfair's proposal.

Guthrie was fascinated by the chemistry of a class of hydrocarbons with carbon–carbon double bonds then known as "olefines" but now more commonly called "alkenes." In his experiments that led to the synthesis of mustard gas, Guthrie first separated "ethylene from ethylic alcohol" as he put it. The modern names for these compounds are respectively ethene and ethanol, although the old names, ethylene and ethyl alcohol, are still used. Guthrie then bubbled ethene (a gas) at the rate of about one bubble a second through each of the two sulfur chloride compounds, both of which are liquids. He described his findings in a paper published in 1860.[10]

Guthrie misleadingly called the two sulfur chloride compounds "bisulphide of chlorine" and "chloride of sulphur." By the time of the First World War, when Pope came on the scene, the two compounds were known as "sulphur monochloride" and "sulphur dichloride." To make matters even more confusing, the modern convention, as recommended by the International Union of Pure and Applied Chemistry (IUPAC), is to use the spellings "sulfur" and "sulfide" rather than "sulphur" and "sulphide." In this book, we use the "f-spelling" except when quoting directly from historic documents when we use the "ph-spelling."

In the first experiment, on sulfur monochloride, Guthrie had little success. "Dry ethylene may be passed for hours through bisulphide of chlorine at temperatures varying from 0 °C to 100 °C without the occurrence of any appreciable chemical change," he wrote. The story was different with sulfur dichloride, however. The colour of the liquid changed from "garnet-red" to "straw-yellow" as the ethene bubbled through. The product of the reaction was an oily liquid which he called "bichlorosulphide of ethylene." Guthrie not only sniffed the vapour that the liquid gave off, he also tasted the liquid and tested it on his skin!

"Its smell is pungent and not unpleasant, resembling that of oil of mustard; its taste is astringent and similar to that of horse-radish," he wrote. "The small quantities of vapour which it diffuses attack the thin parts of the skin, as between the fingers and around the eyes, destroying the epidermis. If allowed to remain in the liquid form on the skin, it raises a blister." In other words, it was a vesicant.

Pope and his co-worker, British chemist Charles Gibson (1884–1950), concluded that the liquid product with a vesicant action (which Guthrie had called "bichlorosulphide of ethylene") was "doubtless a highly impure" dichloroethyl sulfide (that is mustard gas).[11] The two chemists pointed out that "this observation was shortly afterwards confirmed" by German chemist Albert Niemann (1834–1861) in a paper describing similar research that he had carried out independently.[12] The previous year, Niemann had isolated cocaine from coca leaves. He died in 1861 at the age of 26 from lung problems, possibly caused by inhaling vapours given off by mustard gas.

Guthrie conducted further experiments on the reaction of ethene with sulfur monochloride, this time immersing a "well-stoppered bottle" containing the two compounds in boiling water for 20 hours. He repeated the procedure several times, each time refilling the bottle with ethene. Finally, after a fairly elaborate purification process, he obtained a pale yellow liquid with vesicant properties. He published his findings in 1861.[13] The paper revealed that he had once again tasted and

Figure 14.2   Charles Gibson (Royal Society of Chemistry Library).

smelled the toxic liquid. It had "a not unpleasant but in-describable smell; its taste is intensely sweet and pungent," he wrote adding, "its annoying effect upon the eyelids is very enduring."

But whereas Niemann survived just a year or so after his contact with mustard gas, Guthrie lived on eventually becoming professor of physics at the Royal College of Science in London. The Physical Society of London, which eventually became part of what is now the Institute of Physics, was formed on his initiative in 1874. He was president of the society from 1884 until his death in 1886, six days after his 53rd birthday.

THE MEYER–CLARKE ROUTE

In the same year that Guthrie died, Viktor Meyer (1848–1897), a German organic chemist, synthesised mustard gas by a relatively complicated route that was completely different from that of

Guthrie's apart from the initial step.[14] Like Guthrie, he first prepared ethene from ethanol. He then passed the ethene and carbon dioxide through a solution of bleaching powder (which consists mainly of calcium hypochlorite). The product of the reaction was a chlorine-containing organic compound, known as 2-chloroethanol or ethylene chlorohydrin, which has a molecular structure similar to that of ethanol. He then treated the product with sodium sulfide to form a sulfur-containing organic compound and finally converted this compound into mustard gas by treating it with phosphorus trichloride. Meyer soon realised the extremely toxic nature of mustard gas and decided to discontinue his work on the compound.[9]

Meyer, who was professor of chemistry at the University of Göttingen when he published his method of preparing mustard gas, was a workaholic and an insomniac. He suffered from ill health including depression for many years and finally committed suicide by taking cyanide at the age of 49. His obituary in the *Journal of the American Chemical Society* outlines his many contributions to organic chemistry but does not refer to his work on mustard gas, maybe because at the time it was not considered to be too important.[15]

Two years before the start of the First World War, Hans Thacher Clarke (1887–1972), a chemist who was born in England but had an American father and German mother, showed that very high yields of mustard gas could be rapidly obtained in the final step of Meyer's process if the phosphorus trichloride was replaced by concentrated hydrochloric acid.[16] Like Meyer, he observed that dichloroethyl sulfide is intensely poisonous and inflicted painful wounds which only healed with the greatest difficulty. Clarke referred to "the absence of pungency in the odour" and noted that "it takes effect only after some hours have elapsed." But unlike Guthrie, he did not taste the liquid, or if he did, he did not mention it in the paper reporting his work.

Clarke carried out his research on mustard gas while working in the laboratory of German chemistry professor and Nobel Prize winner Emil Fischer at the University of Berlin (see Chapters 4 and 18). German theoretical physicist Max Planck (1858–1947), who was awarded the 1918 Nobel Prize in Physics for his research on quantum theory, was professor of theoretical physics at the university at the same time. Clarke married Planck's niece Frieda Planck (1890–1960) in 1914.[17]

With war looming and the prospect that Germany would not be able to export chemicals to the United States, George Eastman (1854–1932), the American entrepreneur who had founded the Eastman Kodak Company, invited Clarke to move to Rochester, New York, to consult and carry out research on the photographic chemicals used to make roll films. The newly married couple emigrated to America and Hans spent the war years working for Eastman's company. He stayed there until 1928 when he took up an appointment as professor of biological chemistry and head of department at the Columbia University College of Physicians and Surgeons.

Guthrie, Meyer and Clarke were all civilian chemists. Their papers describing methods of preparing mustard gas all referred to the toxicity and dangers of the compound but none of them alluded to its potential applications let alone its possible military exploitation in warfare. Their focus was on chemistry rather than on its uses.

Even so, the idea of using mustard gas as a chemical weapon can be attributed to Clarke, albeit indirectly. While experimenting with the compound in Fischer's laboratory in Berlin, a flask broke splashing some of the corrosive oily liquid onto his leg.[9] He suffered severe burns and was hospitalised for two months. Fischer dutifully reported the accident to the German Chemical Society in 1915 and as a consequence, Clarke suggested, the Germans began to investigate its potential for chemical warfare. The German chemical company Bayer subsequently used the Meyer–Clarke method to produce the liquid, adapting its dye-manufacturing works at Leverkusen to do so.

HOT AND COLD PROCESSES

After the introduction of mustard gas on the battlefield by the Germans in July 1917, British and French chemists analysed samples of the toxic liquid and rapidly concluded that it was dichloroethyl sulfide. As it proved so effective as a chemical warfare agent, the Allies decided to use it themselves. But first they needed to develop their own procedures for its manufacture.

The British initially asked Scottish organic chemist James Irvine (1877–1952) to develop a process for making the agent. Irvine, a professor of chemistry at the University of St Andrews, and co-workers decided to explore the Meyer–Clarke route but were unable to produce the chemical in high yield. Irvine was also concerned that the raw materials needed to make the compound would not be available in sufficient quantities. The St Andrews team then began to study the Guthrie process but by September 1917 once again realised that they did not have enough of the precursor chemicals to continue.[18]

The following month, Pope and co-worker Gibson were invited to tackle the problem. Pope was not impressed by the Meyer–Clarke method or the way the Germans manufactured the toxic chemical. "Most British organic chemists were, I think, amazed at the method of production adopted by the German manufacturers; to apply such a technically cumbrous process for the manufacture of so simple a compound seemed quite irrational," he remarked in his 1919 presidential address to the Chemical Society and then added scathingly: "The German chemical service was inefficient; the scientific chemists under its control were incompetent."[1]

Gibson and Pope decided to focus on Guthrie's process. "In the late autumn of 1917 we were requested by the Chemical Warfare Department to study the preparation of [mustard gas] and in this connexion we thought it desirable to investigate carefully the interactions of ethylene and the two chlorides of sulphur," they wrote in their paper, published in 1920, that reported their work on mustard gas.[11]

The paper describes how the two chemists managed to develop Guthrie's process for making mustard gas in very high yields from sulfur monochloride and ethene at temperatures between 50 °C and 70 °C. It became known as the "hot" process. The two chemists reported the method to the Chemical Warfare Department in early 1918. "The new method was communicated to France and America, and installed by the three Great Allies on a large scale;" Pope informed the Chemical Society in his 1919 presidential address,[1] adding: "at the conclusion of the armistice the available daily production of mustard gas by the Allies was equal to the monthly production of the Central Nations." The Central Nations were Germany and its allies.

French chemists also worked on the development of Guthrie's monochloride process and discovered a way of making mustard gas at 30 °C to 40 °C, the so-called "cold" process, which proved to be superior to the "hot" process. The latter apparently produced "weak, ineffective batches of mustard gas."[19]

In England in April 1918, the Manchester-based dye-manufacturing company Levinstein began to adapt the "cold" process to produce small amounts of good quality mustard gas. The firm licensed their process to the Americans who then used what became known as the Levinstein process to make mustard gas at their Edgewood site in Maryland.

The manufacture of mustard gas by the Guthrie process depended on supplies of sulfur monochloride and ethene. Sulfur monochloride is prepared by the reaction of sulfur and chlorine. Sulfur, a yellow solid, was available in abundance in underground and volcanic deposits around the world at the time of the war. Chlorine was extracted from brine (a solution of sodium chloride in water) by electrolysis.

Ethene was a key starting material for both the Guthrie and Meyer–Clarke processes. The Germans obtained the gas from ethanol which was readily produced by fermentation. Their process split the ethanol into ethene and water. It involved heating the alcohol with an aluminium oxide catalyst at 400 °C. The Allies investigated other ways of producing the gas such were the demands for it in 1918. They included cracking crude oil and extracting it from coke oven gas.[20]

THE MOST EFFECTIVE OF BATTLE GASES

Mustard gas was "by far the most effective" of the battle gases used in the First World War concluded Augustin Prentiss in his book on chemicals in war.[21] Unlike chlorine, phosgene and the other lethal agents used earlier in the war, it did not blow away in the wind. It was persistent and could lie around on the ground and in trenches for days on end, especially in cold weather.

Prentiss, a lieutenant colonel in the Chemical Warfare Service of the United States Army, calculated that on average just 60 pounds of mustard gas were required to produce one casualty. That compared with 500 pounds of high explosive for each high explosive casualty and 5000 rounds of rifle or machine-gun ammunition for each bullet-wound casualty.

From July 1917, when the German artillery first fired mustard gas shells, until the end of the war, mustard gas accounted for some 125 000 British casualties, which is approximately 70% of all British gas casualties in the war, according to one source,[22] although another source puts the percentage at 90.[23] But even though the gassing rarely proved fatal, just 1.5% of the casualties died, it often inflicted terrible long-term injuries including permanent blindness.

The Germans used the codename "Lost" for the chemical. The name derived from the first two letters of the surnames of the two chemists who developed the process to mass produce the liquid for use in the war. They were Wilhelm Lommel, who worked for Bayer, and Wilhelm Streinkopf (1879–1949), who headed the chemical weapons research team at the Kaiser Wilhelm Institute for Physical Chemistry and Electrochemistry in Berlin. The Germans also called mustard gas "Yellow Cross" after the markings on the gas shell.

The French codenamed the vesicant "yperite" because it was first employed at Ypres in Belgium. The British dubbed it "HS" which was short for "Hun Stuff" and together with the Americans also called it "mustard gas" because its smell was similar to that of mustard. The smell was caused by impurities in the liquid. When pure, it has little or no smell. By the end of September 1918, Britain alone was delivering in excess of 90 000 HS shells each week to the Royal Artillery on the Western Front.[24]

Manufacture of the lethal gas was technically difficult and also dangerous. The French were producing mustard gas and firing shells with the highly toxic chemical by June 1918, but "casualties multiplied at the works," noted British chemist Victor Lefebure who was British Chemical Warfare Liaison Officer with the French at the time.[25] Lefebure refers to a French minister visiting a factory and "pinning the Legion of Honour on to the breast of a worker blinded by yperite."

After American's entry into the war in 1917, the United States Army inducted American chemist James Conant (1893–1978) into its Chemical Warfare Service and put him charge of the technical development of mustard gas. Conant had received his PhD in organic chemistry from Harvard University in the

Figure 14.3   James B. Conant (William Haynes Portrait Collection, Chemical
Heritage Foundation Image Archives, Philadelphia, PA).

summer of 1915 and, impressed by the achievements of German
chemists, had hoped to travel to Germany to continue his research there. But even though the United States had yet to enter
the war, that proved impossible.[26]

Conant's team developed the Levinstein process based on the
"cold" version of Guthrie's method to make mustard gas at the
Edgewood site. From August 1918 onwards the plant produced
an average of ten tons of the toxic liquid each day.[27] Conant was
also assigned responsibility for the development and manufacture of Lewisite, an even more toxic vesicant, at a factory in
Cleveland, Ohio. The armistice was signed, however, before the
arsenic-containing organic chemical could be used.

By the time the war ended, Germany, France, Britain and the
USA had produced a total of around 11 000 tons of mustard gas.
Germany produced 7600 tons, which was about 8% of its total
production of poison gases during the war.[28] France, Britain and

the USA accounted for 2000 tons, 500 tons and 900 tons of the total, respectively.[29] If the war had continued beyond November 1918, the Allies' output of the agent would by far have exceeded that of Germany. According to British historian of science Jack Morrell, Britain alone was producing 30 times more mustard gas than Germany at one thirtieth of the cost in the final months of the war.[30]

In Italy, a team of chemists working in a laboratory at the University of Bologna to develop processes for making the chemical agent had, by the end of September 1918, prepared some samples of the toxic oily liquid. It is doubtful whether Russia managed to produce the agent during the war.[31]

AFTER THE WAR

How times have changed! Nowadays mustard gas is anathema. The very existence of the lethal liquid and other chemical warfare agents, let alone their development and production, is regarded as an affront to the civilised world. The manufacture, stock-piling, and use of such chemicals are banned under the Chemical Weapons Convention administered by the inter-governmental Organisation for the Prohibition of Chemical Weapons. The organisation, which has its headquarters in The Hague, Netherlands, won the Nobel Peace Prize in 2013 "for its extensive efforts to eliminate chemical weapons."

Yet after the First World War, Britain, France, and the United States eagerly recruited, promoted, lauded and honoured the chemists who had contributed to the development and pro-duction of mustard gas and other lethal chemicals used by the military in the war.

Conant, for example, who had risen from the rank of first lieutenant to major while in the United States Army, received several job offers soon after the end of war as a result of his "outstanding work" on mustard gas and Lewisite.[26] He opted to return to Harvard University, however, and in 1919 took up an appointment as assistant professor of chemistry there. He be-came professor of organic chemistry at the university in 1929. Four years later Conant accepted an offer to become its presi-dent. In 1953, he resigned the Harvard presidency and, at the invitation of United States president-elect Dwight D. Eisenhower

(1890–1969), became High Commissioner for the United States Zone of Occupation in Germany. He subsequently became the first United States ambassador to the Federal Republic of Germany (West Germany).

Conant received numerous honours and awards after the war. In 1936, for instance, the French made him a commander in its Legion d'Honneur. In Britain, he was elected a foreign member of the Royal Society and honorary fellow of the Royal Institute of Chemistry. In 1948, he was made an honorary commander of the Order of the British Empire (OBE).

GIBSON AND POPE ALSO HONOURED

Charles Gibson, Pope's wartime co-worker on mustard gas, was awarded the OBE in 1919 in recognition of his services to the country during the war. After graduation from Oxford University with an honours degree in chemistry in 1906, Gibson had worked as a research assistant with Pope in Manchester and then as a lecturer and demonstrator at Cambridge. In 1912, he moved to India to take up an appointment as professor of chemistry at Maharajah's College at Trivandum, a college affiliated to the University of Madras.

Following the outbreak of war, Gibson returned to England and rejoined Pope in Cambridge. He was honorary adviser to the Chemical Warfare Committee of the Ministry of Munitions from 1916 to 1919. While working with Pope on the manufacture of mustard gas, Gibson suffered gas poisoning following an accident and "there is little doubt that he never fully recovered from this," according to organic chemist John Simonsen (1884–1957) in his obituary of Gibson.[32] In 1921, after a couple of years as professor of chemistry at the Egyptian School of Medicine, Cairo, Gibson was appointed to the chair of chemistry at Guy's Hospital Medical School, University of London.

Simonsen, like Gibson, was born in Manchester in 1885. They both attended Manchester Grammar School and both were professors at colleges in Madras in the years before the war. During the war, Simonsen remained in India as controller of oils and chemical adviser to the Indian Munitions Board. In 1928, he returned to England and spent a couple of years working with Gibson at Guy's before becoming professor of chemistry in the

British Association Meeting, Dublin, 1908
John Simonsen and Robert Robinson in Camp near Howth

Figure 14.4 John Simonsen (left) and English organic chemist Robert Robinson (1886–1975) in camp near Howth, during the 78th meeting of British Association for the Advancement of Science, held in Dublin, 1908. Robinson won the 1947 Nobel Prize for Chemistry for his research on alkaloids, such as morphine, and other plant natural products (Royal Society of Chemistry Library).

University of Wales at University College, Bangor. Simonsen was knighted in 1949.

Pope was honoured with a knighthood 30 years earlier for his services during the First World War. Sir William Jackson Pope, as he now was, continued to have strong views about the Great War and its conduct and was particularly scathing about what he called the "military men." War was always a "very dirty business," he observed in an address on "modern developments in war making" that he delivered at the 62nd meeting of the American Chemical Society held at Columbia University in September 1921.[33] The modern battlefield, he said, "was one of the foulest places that one can imagine."

The universal extolling of the chivalry associated with war did "a very bad service" to humanity, he added. Pope was no doubt alluding to the view of David Lloyd George (1863–1945), who was British Prime minister at the time, and other leading British figures that Britain's part and conduct in the war had been both chivalrous and honourable.

Pope considered that the military's conservatism and opposition to innovation stimulated and prolonged war. "The military mind always resents anything that is new. That has been the case throughout the whole of history." Then, paradoxically, he described three "innovations in actual weapons" that the military introduced during the First World War. The first, which he just mentioned in passing, was the aeroplane. The second was chemical warfare. As we saw in the second paragraph of this chapter, he considered toxic agents such as mustard gas to be "a very humane weapon of war." He added: "Public opinion on this subject is grotesquely ill-informed, at any rate in England."

The third innovative weapon he mentioned was, curiously, preventive medicine. "Every great soldier has realized that an army is limited in size by the difficulty of keeping it free from disease and epidemic in the field," he remarked. "Preventive medicine made it possible to maintain twenty million men under arms and abnormally free from disease, and so provided greater scope for the killing activities of the other powers."

Pope was vice-president of the International Union of Pure and Applied Chemistry (IUPAC) when he gave his address. The union, formed on 28 July 1919, represents chemists of its member countries. It promotes cooperation among chemists around the world and also seeks to contribute to the advancement of pure and applied chemistry "in all its aspects."

IUPAC was set up following the dissolution of the International Association of Chemical Societies (IACS). The association had held its inaugural meeting in April 1911. The founder members were the British, French and German chemical societies. By the following year, ten other societies had joined the association including those of Austria, Italy, Russia and the United States. But then war broke out. The association "never recovered from this catastrophe" and was "forced to cease functioning" notes Roger Fennell, author of the union's history.[34]

In 1923, Pope became the second president of the union. He took over from another eminent chemist who had also taken a leading role in chemical warfare research and development and mustard gas production during the First World War.

THE FIRST IUPAC PRESIDENT

The first president of IUPAC was French organic chemist Charles Moureu (1863–1929). He held the office from 1920 to 1922. Moureu and Pope were key figures among a group of European and American chemists charged with establishing the union in 1919.

Born in Mourenx in the Aquitaine region of southwest France, Moureu had a distinguished career in chemistry culminating in his appointment as professor of organic chemistry at the prestigious Collège de France in Paris in 1917. By then he had also become heavily involved in the French chemical warfare effort.

The French embarked on chemical warfare, albeit non-lethal chemical warfare, at the start of the war in August 1914 when they filled rifle grenades with ethyl bromoacetate. The organic chemical was a type of chemical warfare agent known as a lachrymator or more commonly as a tear gas. In March 1915, their artillery began firing gas shells filled with the lachrymator and subsequently other lachrymatory halogen-containing organic chemicals, notably chloroacetone, benzyl bromide, benzyl iodide and iodoacetone.[35]

A few days after the lethal German chlorine gas assault at Langemarck in April 1915, French military representatives and chemists met and established a commission to develop a chemical warfare programme. By August, academic chemists in 13 laboratories around the country were carrying out research on and development of chemical warfare. Two teams were established. One focused on the offensive use of chemical weapons and the other on defensive aspects, especially the development of gas masks and other forms of protection. Moureu led the offensive chemical warfare team.

Moureu's "brilliant organic investigations characterised later French developments," commented Lefebure.[36] The French began to employ gas shells filled with phosgene, a lethal choking agent, at the start of the Battle of Verdun which lasted from 21 February to 18 December 1916. And then, in July 1917, the Germans fired their first mustard gas shells. The French were impressed with this new chemical weapon and decided to look into the possibility of developing their own programme for its production.

Moureu and his fellow chemists initially examined the Meyer–Clarke process that the Germans were using but made little progress for a variety of reasons: lack of experience working on such as process, technical difficulties, and limited supplies of the chemicals needed to make mustard gas by this route. The chemists therefore turned their attention to Guthrie's method and soon developed the so-called "cold" sulfur monochloride process for manufacturing the chemical.

The French started producing mustard gas in March 1918 and filling shells with it the following month. Their artillery began firing the shells on the Western Front in June 1918, some three months before the first British mustard gas shells were fired at the front. According to Prentiss, the French filled over almost 2.5 million shells with the vesicant in 1918. The shells "proved very effective in battle during the last six months of the war," he comments.[37]

POSITIVE CONTRIBUTIONS TO CHEMISTRY

Conant, Gibson, Pope, and Moureu were all accomplished chemists who made significant contributions not only to the war effort but also to the advancement of other aspects of pure and applied chemistry throughout their careers. After the war, for example, Conant achieved an international reputation for his research on physical-organic chemistry and the chemistry of natural products such as chlorophyll and haemoglobin.[38]

Gibson wrote numerous papers describing research on natural products, stereochemistry, organoarsenic chemistry and, most importantly organic compounds containing gold.[32] Most of these papers were co-authored with Pope, Simonsen and other colleagues. Gibson also wrote several books, including one on the principles of organic chemistry.[39]

Pope's research spanned a wide range of topics in chemistry although he is best known for his research on optical activity.

His discoveries of optically active compounds of nitrogen, sulfur, selenium and tin, for example, "will always remain landmarks in the history of the science," according to Mills.[40] He carried out much of this research before the war. "After the war Pope became more and more involved in the administration of chemical affairs," Mills notes.[41]

Moureu was not only a distinguished chemist, he was also a prolific author before and after World War I. His book *Fundamental Principles of Organic Chemistry* was published in French in 1902 and in English in 1921.[42] Another book of his, *The Chemistry and the War: Science and Future*, was published in French in 1920.[43] The book has chapters on chemistry and metallurgy, explosives, combat gases, chemistry and aeronautics, chemistry and camouflage, and chemistry and health.

STUDY OF WAR GASES YIELDS BENEFITS

Research on chemical warfare agents during the war surprisingly resulted in some beneficial outcomes. Moureu's research on acrolein is one of the most significant examples. The organic chemical, also known as propenal or acrylic aldehyde, is a flammable liquid with a pungent smell. It is formed by the thermal decomposition of glycerine. Before the war broke out, Scottish chemist William Ramsay carried out experiments on acrolein and subsequently suggested to the British War Office that the chemical could be used as a tear gas against the enemy. The office rejected his proposal.

The French, however, introduced acrolein in January 1916 as a filling for artillery gas shells and gas grenades. Although the chemical is a potent lachrymator and respiratory irritant, it proved to be unsuccessful as a chemical warfare agent owing to its lack of chemical stability. Moureu and his co-worker, French organic chemist Charles Dufraisse (1885–1969), observed that when it was exposed to air, it spontaneously turned into a harmless waxy resin.

The two French chemists subsequently proved that the deterioration of the acrolein could be attributed to atmospheric oxygen and a process known as autoxidation.[44] They showed that oxygen and acrolein combined to generate peroxides that acted as a catalyst and therefore hastened the rate of deterioration.

Peroxides are compounds consisting of molecules containing two oxygen atoms linked together by a single bond.

Moureu and Dufraisse called the compounds "pro-oxygenic."[45] They also demonstrated that the decomposition of acrolein could be retarded by "anti-oxygenic" compounds, most notably phenolic compounds. These compounds consist of molecules with benzene rings bonded to one or more hydroxyl groups.

Nowadays, "anti-oxygenic" compounds are known as "anti-oxidants." They have a wide range of applications as additives that inhibit the oxidation and consequent degradation of materials. For instance, they are added to rubber to slow ageing and to fats to stop them going rancid. Antioxidants are also ubiquitous in nature. Vitamin C, also known as ascorbic acid, is just one example. The antioxidant is present in citrus fruits, green vegetables, potatoes and other fruits and vegetables. The vitamin maintains healthy skin and helps cuts and abrasions to heal properly.

Our current knowledge of antioxidants and how they contribute to health and prevent disease can therefore be traced back to research on a chemical warfare agent, acrolein, by a French chemist who helped to develop an efficient process for making mustard gas and later became the first president of IUPAC.

THE BARI MUSTARD GAS TRAGEDY

The research, development, production and use of mustard gas in the First World War had terrible consequences with thousands of men killed by the chemical and many more permanently blinded or injured in one way or another. But there was one beneficial outcome from the use of the agent, although it would take a tragedy some 25 years later to uncover it fully. And for that, we need to move on to the Second World War and a German air raid on Bari on 2 December 1943.

Bari is a port and city in the south of Italy on the Adriatic coast. During World War II, the port became an important hub for the Allies following their invasion of mainland Italy in September 1943. By the start of December, Bari harbour had become crowded with ships, mainly merchant cargo vessels, unloading ammunition and essential supplies for Allied troops who were advancing towards Rome.

On the evening of 2 December, several German aircraft flew over the harbour dropping pieces of aluminium foil to confuse the Allies' radar defences. Over 100 German Junker bombers then carried out a surprise attack dropping bombs on the ships. The raid, which lasted less than an hour, destroyed 27 ships and another dozen were damaged. They included several British ships and eight of the thousands of cargo ships hastily built in the United States during the war known as "Liberty ships." Two of the Liberty ships that sank, the SS (steamship) *John Harvey* and the SS *John Motley* were loaded with ammunition. Direct bomb hits on some of the ships caused fires and massive explosions that ruptured oil pipelines and shattered windows miles away.

The *John Harvey* was carrying a highly secret cargo of 2000 chemical bombs designed for aerial bombardment. Each bomb was filled with some 30 kg of mustard gas. The use of mustard gas and other chemical weapons was prohibited by the Geneva Protocol which had been signed in 1925 and came into force in 1928. The signatories included France, Germany, Italy, the United Kingdom and the United States. The protocol, however, did not deter the Allies from manufacturing mustard gas and planning to use it in response to any German poison gas attack in Italy. Germany produced tens of thousands of tons of chemical warfare agents during the Second World War. They included mustard agents and organophosphorus nerve agents such as tabun and sarin. German dictator Adolf Hitler, however, did not give the order to use them. If he had done so, the weapons might well have altered the course and outcome of the war.

At the time of the Bari attack, the *John Harvey* was moored in the harbour waiting to unload the chemical weapons. The port authorities were not aware of the poisonous cargo and had therefore not given the ship priority to unload. Although it did not receive a direct bomb hit, it caught fire from burning debris thrown up from other ships that had been hit. It then exploded killing everyone on board. The fire and explosion destroyed most of the mustard gas but a substantial quantity also spattered into the harbour and mixed with fuel oil escaping from the wrecked ships and damaged pipelines. To make matters worse, the toxic agent began to evaporate and the vapour quickly mingled with the clouds of smoke above the water.

Many of the survivors from the other ships blasted by the bombs were blown into the water and swam for their lives through the slicks of mustard gas and oil. As they did so they also inhaled the toxic vapours. Some managed to make it to life rafts, a number of which had become drenched with sea water contaminated with the poisonous oil. They clung to the sides of the rafts while others clambered into them and sat there with oily water swishing around their feet. Inevitably the oily mixture seeped through their clothing and into their shoes.

The medical officers and nurses at the local hospitals were overwhelmed by the number of casualties. Because they were not aware of the deadly cargo in the harbour and as the fuel oil, smoke and sea water obscured the faint smell of the gas, the hospital staff did not realise that many of the sailors rescued in the raid had been exposed to the mustard agent. They provided the survivors with warm tea, wrapped them in blankets, and left them for up to 24 hours with their clothing and bodies soaked in the poisonous cocktail. Meanwhile, they tended to what they considered to be the more urgent cases: the seamen who had suffered severe blast wounds, burns, and other types of injuries during the raid.

Mustard gas has a delayed effect. It took several hours before survivors of the Bari raid started to display symptoms of exposure to the agent: first reddening of the skin and then burns, blisters, breathing difficulties and eventually blindness. A few of the sailors died the following morning, within 24 hours of the raid. The number of deaths peaked on the second and third days after the attack and then slowly decreased over the following ten days or so. Some of the medical staff who treated the victims and handled their contaminated clothing also became casualties of mustard gas poisoning, sustaining skin burns and irritation of the eyes. Furthermore, many citizens also died after inhaling the gas that had wafted over the city following the raid. Others came into contact with the contaminated sea water and oil that had splashed over the harbour walls.

Estimates of the number of casualties and fatalities caused by the attack vary. According to one account, the bombing killed one thousand servicemen and another thousand civilians.[46] Over

800 casualties were hospitalised with burns and other injuries and wounds caused by the blasts of the explosions, the fires, and the mustard gas. Around 70 of the hospital deaths in the first two weeks after the raid were attributed wholly or partly to mustard gas poisoning.

THE COVER UP

General Dwight D. Eisenhower (1890–1969), who was in charge of the Allied invasion of Italy at the time of the Bari attack, had just been appointed supreme commander of the Allied expeditionary force in Europe. Following the raid, he connived with the president of the United States, Franklin D. Roosevelt (1882–1945), and British Prime Minister Winston Churchill to cover up the mustard gas disaster. However, as news of the tragedy started to leak out over the following weeks, the Allies issued a statement acknowledging that a chemical weapons accident had occurred but confirmed that they had no intention of using poison gas unless the Germans did so first. In order to keep the mustard gas poisonings as secret as possible, many records were destroyed and deaths resulting from the poisoning were attributed to burns and other causes.

Even though official United States records of the raid were released in 1959, there was little interest in the catastrophe until *Disaster at Bari* was published in 1971. The disaster was "one of the best kept secrets of World War II," the author, Glenn B. Infield, observed in the book.[47] Even in 1961, he found that "few military officials knew or would admit that there had been any mustard at Bari."[48] He added: "How many eventually died or were disabled by their exposure to the mustard at Bari that night will never be known."

Infield was a bomber pilot during the Second World War and subsequently wrote numerous books and articles about the war. He died in 1981 at the age of 60.[49]

Immediately after the Bari tragedy, Stewart F. Alexander (1914–1991), who was a member of the United States Army medical staff in Europe, travelled to Bari to examine the survivors in the hospital wards and investigate the mysterious deaths. Alexander, a doctor and an expert in chemical warfare medicine, had not been aware of the toxic cargo on board the *John Harvey*. However, the unexplained deaths following the German air strike suggested that a chemical weapon may have been deployed. As soon as he arrived, Alexander was taken to the nearest hospital where he immediately detected and recognised the distinctive smell of mustard gas.

Much of his preliminary report on the catastrophe, dated 27 December 1943, was based on statements made by casualties and medical officers and nurses who cared for the victims.[50] Some survivors, he noted, commented on the "garlicky odor" and jokingly attributed it "to the quantities of garlic consumed by the Italians." Alexander soon established that the burns, blisters, and other symptoms of the victims were consistent with mustard gas poisoning. He concluded that "the mixture or solution of mustard in oil ... produced most of the severe casualties and deaths."

THE BIRTH OF CANCER CHEMOTHERAPY

Tissue samples taken during the autopsies of some of the victims who died at Bari revealed very low levels of lymphocytes, a type of white blood cell in the immune system that helps to protect the body against infectious diseases. The investigations also showed that cells in the blood-forming tissue of the victims' bone marrow were depleted. The condition is known as "bone marrow aplasia."

These observations were not new. In 1919, researchers reported that mustard gas impaired the functioning of bone marrow. The findings were based on experimental work with rabbits and also on the blood counts of soldiers who had been "gassed" in the First World War.[51]

Scientists soon realised that mustard gas and other "mustards" might be used as anti-cancer drugs to suppress the abnormally rapid division and proliferation of blood and bone marrow cells. The abnormal growth of these cells leads to cancers such as leukaemia, a disease that interferes with the production of normal blood cells, and lymphomas, malignant tumours of the lymph nodes. In the years running up to the Second World War, a number of research groups carried out various studies of the biochemistry and pharmacology of "mustards."[52] The term "mustards" in this context refers to mustard gas, other organic compounds with similar chemical structures generally known as "sulfur mustards," and the nitrogen mustards.

Nitrogen mustards are a group of organic compounds known as chloroalkylamines. They were first synthesized by several groups of scientists in 1935.[53] The compounds are chemically similar to sulfur mustards, the major difference being the presence of nitrogen atoms instead of sulfur atoms in their molecules. Nitrogen mustards typically have a faint fishy odour and are strong vesicants. Three nitrogen mustards were produced for use as chemical weapons in the Second World War. Although they were stockpiled, they were never used.

Early in 1942, Alfred Gilman (1908–1984) and Louis Goodman (1906–2000), two prominent pharmacologists at Yale University, United States, were asked to look into the possible clinical use of nitrogen mustards as chemotherapeutic agents. The work formed part of an American classified wartime programme to investigate the therapeutic effects of chemical warfare agents. The two pharmacologists soon observed that the nitrogen mustards were not only vesicants but were also cytotoxic, that is, they damaged or destroyed cells. The systemic effects of nitrogen mustards led to the death of experimental animals "even after topical application to the skin," Gilman noted many years later.[54]

Following the attack at Bari and Alexander's investigations, work on the potential therapeutic use of nitrogen mustards intensified. Gilman and Goodman carried out experiments with mice and demonstrated that one of the nitrogen mustards, known as mustine or mechlorethamine, offered promise for treating lymphomas. In 1943, they persuaded Gustaf Lindskog (1903–2002), a Yale colleague and leading thoracic surgeon to carry out a clinical trial with the chemical.[55] Linkskog administered mustine intravenously to a 47-year-old patient with advanced lymphosarcoma. The dying man willingly agreed to the treatment. Months of radiation treatment had failed to halt the progress of the cancer. He was unable to move his head, eat or speak. Within a few days of the nitrogen mustard injection, the large tumours in his head and neck reduced in size significantly but then returned when the treatment was withdrawn. A second dose had less impact on the tumours and a third dose no effect at all. The cancer cells had developed resistance to mustine.

Gilman and Goodman conducted further trials on the use of nitrogen mustard in "the treatment of certain diseases of the blood-forming organs." The trials were carried out in secrecy and not reported until 1946, the year after the end of the Second World War. "The chemical, pharmacologic, toxicologic and animal experimental aspects of these compounds" were reviewed in a landmark article by Goodman, Gilman and co-workers the same year.[56]

The nitrogen mustard therapy was not the first example of chemotherapy. German scientist Paul Ehrlich (1854–1915) is regarded as the founder of the science of chemotherapy. Years before the start of the First World War, he showed that drugs could be chemically synthesised to treat diseases. In 1908, Ehrlich and Russian scientist Ilya Ilyich Mechnikov (1845–1916) jointly won the Nobel Prize in Physiology or Medicine "in recognition of their work on immunity." The following year, Ehrlich discovered an arsenic-containing organic chemical that proved highly effective for treating syphilis when administered by intravenous injection. The drug was first marketed under the trade name Salvarsan in 1910.

Nitrogen mustard therapy was, however, the first example of cancer chemotherapy. The work by Gilman, Goodman, Alexander, Linskog and others during the Second World War on mustard agents sparked an explosion of interest in and research on the use of cytotoxic chemicals to cure patients with cancer. Nowadays, "cancer chemotherapy" is simply known as "chemotherapy." Since the Second World War, it has become a well-established and widely known treatment for cancer. The fact that its origins can be traced back to the use of mustard gas as a chemical warfare agent in the First World War is less well known.

REFERENCES

1. W. J. Pope, Presidential Address, *J. Chem. Soc., Trans.*, 1919, **115**, 397.
2. K. C. Bailey, G. T. Morgan, W. Wardlaw, G. T. Moody and W. H. Mills, Obituary notices: James Bell, 1899–1941; Sir Gilbert Morgan, 1870–1940; Sir William J. Pope, 1870–1939, *J. Chem. Soc.*, 1941, 689.

3. Z. Wang, The First World War, academic science, and the 'two cultures': educational reforms at the University of Cambridge, *Minerva*, 1995, **33**, 107.
4. L. F. Haber, *The Poisonous Cloud, Chemical Warfare in the First World War*, Clarendon Press, Oxford, 1986, p. 146.
5. L. F. Haber, *The Poisonous Cloud, Chemical Warfare in the First World War*, Clarendon Press, Oxford, 1986, p. 147.
6. L. F. Haber, *The Poisonous Cloud, Chemical Warfare in the First World War*, Clarendon Press, Oxford, 1986, p. 124 (quote in Haber's book from Pope's letter to Hartley 20 August 1917, Hartley papers).
7. L. F. Haber, *The Poisonous Cloud, Chemical Warfare in the First World War*, Clarendon Press, Oxford, 1986, p. 149.
8. M. Despretz, Des composes triples du chlore, *Ann. Chem. Phys.*, 1822, **21**, 437.
9. R. J. Duchovic and J. A. Vilensky, Mustard gas: its pre-World War I history, *J. Chem. Educ.*, 2007, **84**, 944.
10. F. Guthrie, On some derivatives from the olefines, *Q. J. Chem. Soc.*, 1860, **12**, 109.
11. C. S. Gibson and W. J. Pope, $\beta\beta'$-Dichloroethylsulphide, *J. Chem. Soc., Trans.*, 1920, **117**, 271.
12. A. Niemann, Ueber die Einwirkung des braunen Chlorschwefels auf Elaylgas (The effect of brown chlorosulfide on ethylene gas), *Liebigs Ann. Chem.*, 1860, **113**, 288.
13. F. Guthrie, On some derivatives from the olefines, *Q. J. Chem. Soc.*, 1861, **13**, 129.
14. V. Meyer, Ueber Thiodiglykolverbindungen (About thiodiglycol reactions), *Ber. Dtsch. Chem. Ges.*, 1886, **19**, 3259.
15. G. M. Richardson, Obituary, *J. Am. Chem. Soc.*, 1897, **19**, 918.
16. H. T. Clarke, 3-Alkyl-1 : 4-thiazans, *J. Chem. Soc., Trans.*, 1912, **101**, 1583.
17. H. B. Vickery, Hans Thacher Clarke: 1887–1972. *A biographical memoir*, National Academy of Sciences, Washington, DC, 1975.
18. G. Hartcup, *The War of Invention: Scientific Developments, 1914–18*, Brassey's Defence Publishers, London, 1988, p.108.
19. G. Hartcup, *The War of Invention: Scientific Developments, 1914–18*, Brassey's Defence Publishers, London, 1988, p. 109.
20. L. F. Haber, *The Poisonous Cloud, Chemical Warfare in the First World War*, Clarendon Press, Oxford, 1986, p. 394 (note 99).

21. A. M. Prentiss, *Chemicals in War: A Treatise on Chemical Warfare*, McGraw-Hill, New York, 1937, p. 661.
22. R. Harris and J. Paxman, *A Higher Form of Killing: The Secret History of Chemical and Biological Warfare*, Arrow, London, 2002, p. 27.
23. D. Winter, *Death's Men*, Penguin Books, London, 1979, p. 124.
24. A. Palazzo, *Seeking Victory on the Western Front: The British Army & Chemical Warfare in World War I*, University of Nebraska Press, Lincoln and London, 2000, p. 157.
25. V. Lefebure, *The Riddle of the Rhine, Chemical Warfare in the First World War*, Collins' Clear-Type Press, London and Glasgow, 1921, p. 161.
26. M. D. Saltzman, James Bryant Conant: the making of an iconoclastic chemist, *Bull. Hist. Chem.*, 2003, **28**, 84.
27. L. F. Haber, *The Poisonous Cloud, Chemical Warfare in the First World War*, Clarendon Press, Oxford, 1986, p. 168.
28. W. van Der Kloot, April 1915: Five future Nobel prize-winners inaugurate weapons of mass destruction and the academic-industrial-military complex, *Notes Rec. R. Soc. Lond.*, 2004, **58**, 149.
29. L. F. Haber, *The Poisonous Cloud, Chemical Warfare in the First World War*, Clarendon Press, Oxford, 1986, p. 170.
30. J. Morrell, Research as the thing: Oxford chemistry 1912–1939, in *Chemistry at Oxford: A History from 1600 to 2005*, ed. R. J. P., Williams, J. S. Rowlinson and A. Chapman, Royal Society of Chemistry, Cambridge, 2009, p. 136.
31. G. Pancaldi, Wartime chemistry in Italy: industry, the military, and the professors, and N. M. Brooks, Munitions, the military, and chemistry in Russia, in *Frontline and Factory: Comparative Perspectives on the Chemical Industry at War, 1914–1924*, ed. R. MacLeod and J. A. Johnson, Springer, Dordrecht, 2006, p. 70 and p. 86.
32. J. L. Simonsen, Charles Stanley Gibson. 1884–1950, *Obit. Not. Fell. R. Soc.*, 1950, 7, 114.
33. W. J. Pope, Modern developments in war making, *Ind. Eng. Chem.*, 1921, **13**, 874.
34. R. Fennell, *History of IUPAC 1919–1987*, Blackwell Science, Oxford, 1994, p. 9 and p. 17.

35. A. M. Prentiss, *Chemicals in War: A Treatise on Chemical Warfare*, McGraw-Hill, New York, 1937, p. 441.
36. V. Lefebure, *The Riddle of the Rhine, Chemical Warfare in the First World War*, Collins' Clear-Type Press, London and Glasgow, 1921, p. 95.
37. A. M. Prentiss, *Chemicals in War: A Treatise on Chemical Warfare*, McGraw-Hill, New York, 1937, p. 461.
38. G. B. Kistiakowsky and F. H. Westheimer, James Bryant Conant. 26 March 1893–11 February 1978, *Biogr. Mems Fell. R. Soc.*, 1979, **25**, 209.
39. C. S. Gibson, *Essential Principles of Organic Chemistry*, Cambridge University Press, 1936.
40. W. H. Mills, Sir William J. Pope. 1870–1939, *J. Chem. Soc.*, 1941, 700.
41. W. H. Mills, Obituary: Sir William Jackson Pope, K.B.E., F.R.S, *Analyst*, 1940, **65**, 258.
42. C. Moureu, *Notions Fondamentales de Chimie Organique*, Gauthier-Villars, Paris, 1902; C. Moureu and W. T. K. Braunholtz (translator), *Fundamental Principles of Organic Chemistry*, G. Bell and Sons, London, 1921.
43. C. Moureu, *La Chimie et La Guerre: Science et Avenir*, Masson, Paris, 1920.
44. A. Etienne and R. E. Oesper, Charles Dufraisse, *J. Chem. Educ.*, 1952, **29**, 110.
45. C. Moureu and C. Dufraisse, Catalysis and auto-oxidation, anti-oxygenic and pro-oxygenic activity, *Chem. Rev.*, 1926, **3**, 113.
46. R. Harris and J. Paxman, *A Higher Form of Killing: The Secret History of Chemical and Biological Warfare*, Arrow, London, 2002, p. 122.
47. G. B. Infield, *Disaster at Bari*, Robert Hale & Company, London, 1971, p. xiii.
48. G. B. Infield, *Disaster at Bari*, Robert Hale & Company, London, 1971, p. 248.
49. Anon., Glenn Infield, 60, author of books about aviation and Nazi Germany, *The New York Times*, April 18, 1981.
50. S. F. Alexander, Toxic gas burns sustained in the Bari Harbor catastrophe, Secret report dated 27 December 1943

published in G. B. Infield, *Disaster at Bari*, Robert Hale & Company, London, 1971, p. 258.

51. E. B. Krumbhaar and H. D. Krumbhaar, The blood and bone marrow in yellow cross gas (mustard gas) poisoning, *J. Med. Res.*, 1919, **40**, 497.

52. O. Garai, Nitrogen mustard, *Postgrad. Med. J.*, 1948, **24**, 307.

53. W. E. Hanby and H. N. Rydon, The chemistry of 2-chloro-alkylamines. Part I. Preparation and general reactions, *J. Chem. Soc.*, 1947, 513.

54. A. Gilman, The initial clinical trial of nitrogen mustard, *Am. J. Surg.*, 1963, **105**, 574.

55. V. T. DeVita Jr. and E. Chu, A history of cancer chemotherapy, *Cancer Res.*, 2008, **68**, 8643.

56. L. S. Goodman, M. W. Wintrobe, W. Dameshek, M. J. Goodman, A. Gilman and M. McLennan, Nitrogen mustard therapy, *JAMA, J. Am. Med. Assoc.*, 1946, **32**, 126.

CHAPTER 15

The Biltz Brothers

WILHELM AND HEINRICH

Wilhelm Biltz is most famous for his role as a German tank commander in the Great War, but like his elder brother Heinrich, he also led a distinguished career in chemistry. Wilhelm was born in Berlin on 8 March 1877. Inspired by his brother, he developed a passionate interest in chemistry, particularly in the practical aspects of the subject.

At the age of 23, after earning a doctorate at the University of Greifswald for his research on the chemistry of natural products, Wilhelm moved to the University of Göttingen to take up a post as an assistant to Otto Wallach (1847–1931) who was to win the Nobel Prize in Chemistry in 1910 for his research in organic chemistry. In 1905, Wilhelm was appointed full professor at Clausthal University of Technology where his teaching and research focused on the analysis of inorganic compounds and metals. He was a "hands-on" chemist believing that too much attention was then being paid to chemical theory and not enough to practical work.[1]

His brother, born in Berlin on 26 May 1865, studied chemistry at Berlin University at the time when August Wilhelm von Hofmann, a major figure in the history of German chemistry, was chair of the chemistry department at the university.

The Chemists' War, 1914–1918
By Michael Freemantle
© Freemantle, 2015
Published by the Royal Society of Chemistry, www.rsc.org

Heinrich subsequently obtained his doctorate at the University of Göttingen where another eminent German chemist, Victor Meyer, was professor of chemistry. From 1891 until 1897, Heinrich was professor of chemistry at the University of Griefswald.

He was then appointed professor of inorganic chemistry at the University of Kiel. In 1908, while at the university, Heinrich published the first synthesis of phenytoin (also known as diphenylhydantoin) in the German journal *Chemische Berichte*.[2] Phenytoin is a nitrogen-containing organic compound that many years later was shown to control seizures and was marketed as an anticonvulsant drug to treat major epileptic fits. While carrying out the synthesis, Heinrich worked with his younger brother to co-author a manual on practical inorganic chemistry. The book, *Laboratory Methods of Inorganic Chemistry*, was published in 1909.[3]

Heinrich's research also focused on the chemistry of acetylene and the oxidation of uric acid, an organonitrogen compound which, when it accumulates in excess in the bloodstream, deposits as crystals in the joints causing the extremely painful condition known as gout. He carried out much of this work at the University of Breslau where he was a chemistry professor from 1911 until his retirement in 1933. He died on 29 October 1943.

The Great War interrupted both Heinrich's and Wilhelm's academic careers. Heinrich was 49 years old when the war broke out. He served in the German Army as a reserve officer. Wilhelm held the rank of *"leutnant"* (second lieutenant) in the army throughout the war and was awarded the Iron Cross 1st Class. In April 1918, he commanded one of the tanks that fought British tanks at the Second Battle of Villers-Bretonneux.

TANKS IN WORLD WAR I

During August and early September 1914, the First World War was characterised as a war of movement. German forces invaded Belgium and advanced towards Paris. The French and British put a stop to the advance at the Battle of Marne that took place from 5–12 September 1914. The Germans dug in and static trench warfare on the Western Front rapidly ensued. The front stretched for some 450 miles or more from the North Sea Belgian coast to the Swiss border with France.

The British considered various methods of overcoming the stalemate and within a few weeks came up with the idea of developing "land battleships" with caterpillar tracks that could be used as offensive weapons. These bullet-proof armoured fighting vehicles were intended to resist enemy machine gun fire, flatten barbed wire defences, and cross over trenches, ditches, and other obstacles. In February 1915, Winston Churchill, who at the time was First Lord of the Admiralty, set up a "Landships Committee" to look into the possibility. The committee selected one of the proposed designs and by August that year construction of a prototype known as "Little Willie" was underway in great secrecy at a factory in Lincoln. To conceal its purpose as an armoured fighting vehicle, the factory workers, who were mainly women, were informed that the vehicle was a "tank" to be used for carrying water in the deserts of Mesopotamia.

The prototype vehicle evolved into the British Mark I tank which had its first trial run in January 1916. The tanks first saw action in the First World War on 15 September that year at the Battle of Flers-Courcelette which was part of the Somme Offensive in France fought by the British and French armies against the Germans between 1 July and 18 November. The tanks were classified as either "males" or "females." Each male was armed with four machine guns and two naval guns that fired six-pound shells. The females were designed to carry infantry and armed with five or six machine guns but no naval guns.

Modified versions of the Mark I tank, the Mark II and Mark III, were built in 1916 and 1917 mainly for training purposes. The Mark IV had better armour and improved safety features compared with the earlier models. The Mark IVs, each crewed by eight men, were first used during the Battle of Messines in Flanders, Belgium, that took place from 7 to 14 June 1917. Over 60 tanks supported Allied troops as they advanced and captured Messines Ridge which had previously been held by the Germans. The Mark IVs, however, made only a limited contribution to the successful assault.

British tanks were first used in combat on a mass scale on the first day of the British offensive at Cambrai in north-east France. On the morning of 20 November 1917, some 380 Mark IV tanks supported by almost 100 supply and support vehicles attacked the German front lines. The tanks and accompanying infantry broke through the defences and advanced about four miles on a six mile front. Progress gradually ground to a halt with the loss of 179 tanks. Some were destroyed during the attack and others abandoned when they got stuck in soft ground or failed mechanically. By the end of the day, the British Army had lost approximately 4000 men but also taken over 4000 German prisoners.

The Battle of Cambrai continued until 7 December by which time the Germans had mounted a number of counter attacks and recovered much of the ground they had initially lost to the massed tank attack. Overall, the British suffered around 44 000 casualties and 6000 of their troops were taken prisoner. About 50 000 Germans were killed or wounded and another 11 000 taken prisoner.

The French used combat tanks *en masse* even earlier than the British. On 16 April 1917, the first day of the Second Battle of the

Figure 15.1   The British Mark V tank first saw action against the Germans at the Battle of Hamel in northern France on 4 July 1918 (author's own, 2012, Imperial War Museums, London).

Aisne, the French committed a force of some 130 of its Schneider tanks in a failed attempt to take the Chemin des Dames ridge that was held by the Germans. The ridge lies above the River Aisne in northern France. The battle, which continued until 9 May, was the major component of the Nivelle Offensive conceived and launched by Robert Nivelle (1856–1924), Commander-in-Chief of the French Army. On the first day alone, the French sustained approximately 40 000 casualties and most of the tanks were destroyed or irreparably damaged. By the end of the battle, on 9 May, French and German casualties amounted to an estimated 187 000 and 163 000, respectively. The French failure to break through the German lines and the vast loss of life resulted in widespread mutiny in the French Army. Nivelle was subsequently replaced as commander-in-chief of the army by Philippe Pétain (1856–1951) who employed a variety of measures to restore the morale and improve the fighting effectiveness of the army.

The Schneider tank was armed with a field gun and two machine guns and was relatively light. At 13.5 tons, it weighed less than half the 29 tons of the British male Mark IV tank. The Germans lagged behind the British and French in the development of tanks during World War I. Their A7V tank was developed in 1917 and production began late that year. The tank weighed around 30 tons, was armed with a single cannon and six machine guns, and had a crew of 18.

The A7Vs were introduced at the Battle of St Quentin on the first day of the German attack code-named Operation Michael. The attack marked the beginning of the German Spring Offensive that stretched from 21 March to 18 July 1918. Five tanks saw action at St Quentin, but three broke down and the other two made little impact on the battle.

The A7Vs were in operation again supporting infantry on 24 April in an attack near Villers-Bretonneux, a town about 12 miles east of Amiens in northern France. The attack outside the town was the second attempt by the Germans that month to break through Allied lines and advance towards Amiens. The Germans were unsuccessful in both battles. On the first day of the second attack, three A7Vs, one of them commanded by Wilhelm Biltz, encountered and clashed with three British Mark IV tanks. For the first time in history, tanks fought against tanks.

TANK CHEMISTRY

The first chapter of the Biltz brothers' book on inorganic chemistry focuses on the chemical elements found in the earth's surface, including the oceans and the atmosphere. The brothers point out that the most important naturally occurring compounds include: the oxides and sulfides of elements such as iron, lead and antimony; sodium chloride and other halides; and oxygen-containing salts, for example, nitrates, phosphates, and sulfates. The chapter provides 14 examples of how elements can be extracted from inorganic compounds. The brothers describe, for instance, a laboratory method for preparing lead from lead sulfide, a compound that occurs naturally as the mineral galena. They also show how stibnite, a natural form of antimony sulfide, can be treated in the laboratory to produce antimony.

When their book was published, five years before the start of the Great War, it is almost certain that Wilhelm and Heinrich would have been aware of the important roles iron, lead, antimony and other elements played in the manufacture of weapons, armour, and ammunition. An alloy of lead and antimony, for example, was widely used to make bullets. The two brothers would undoubtedly have been familiar with the explosive properties of inorganic compounds such as ammonium nitrate and potassium chlorate that were to be used extensively in the First World War. And when Wilhelm entered a tank for the first time, he might well have been impressed by the amount of chemistry that went into its construction, its armour, and its gun and ammunition.

In his book about the Second Battle of Villers-Bretonneux, author David R. Higgins provides not only an account of the world's first tank battle and Biltz's involvement, but also detailed technical descriptions of the German and British tanks that fought in the battle.[4] The descriptions inevitably but perhaps surprisingly refer to a variety of chemical materials that were used in the construction and operation of tanks.

The British Mark IV tank, for example, had cast-iron track rollers and employed nickel–steel plate for its armour. Cast iron is a hard and brittle form of iron containing about 2% carbon. Steel, an alloy of iron and carbon containing less than 2% carbon, is stronger than cast iron. The addition of nickel to steel increases the toughness of the alloy. Iron, carbon, and nickel are all chemical elements. The steel plates were hardened by quenching, that is by heating and then rapidly cooling, in water, oil, or molten sodium cyanide. The British and German navies both used a form of nickel steel as armour for their battleships.

The tank's engines had pistons made of another chemical element: aluminium. The Lewis light machine guns that armed both male and female tanks were housed within a phosphor bronze mounting. Phosphor bronze is a strong copper alloy containing about 10% tin and 1% or less of phosphorus. The guns of the male tanks fired six-pound high explosive shells. The steel shell cases were typically packed with amatol, an explosive mixture of ammonium nitrate and TNT.

Brass cartridges attached to the bases of the shells were loaded with cordite, a smokeless propellant made up of nitrocellulose (also known as guncotton), nitroglycerine and petroleum jelly. The brass, another copper alloy, typically consisted of 70% copper and 30% zinc. A percussion primer in the bottom of the cartridge contained a mixture of mercury fulminate and other chemicals that detonated when struck by the firing pin of the gun. The flash from the detonation ignited the propellant which propelled the shell out of the gun. When the shell hit the ground, a trench, or some solid object, the impact detonated a percussion fuse fitted into the nose or base of the shell which in turn detonated the shell's high explosive bursting charge.

The fuse itself required some 50 or more components for its fabrication. They included a copper detonator cap, containing a pellet made of an explosive mixture of chemicals, and a steel needle. A phosphor bronze spring separated the cap and the needle. On impact of the shell, the cap of a base fuse was carried forward onto the needle initiating the detonation. For a nose fuse it was the other way round. The needle was forced into the detonator cap.

The construction of the Mark IV tank, its armour, its guns, and its ammunition therefore all relied on the production and availability of iron, copper, nickel, and zinc and other chemical elements as well as inorganic compounds like ammonium nitrate and nitrogen-containing organic materials such as TNT, nitrocellulose and nitroglycerine.

WILHELM SURVIVES

Just over a month after the start of their Spring Offensive, the Germans attacked Allied lines south of the River Somme in an attempt to reach the city of Amiens, a critical rail hub held by the Allies. The battle began early in the morning of Wednesday 24 April 1918 with a heavy German artillery bombardment of high explosive, gas, and smoke shells. Then, under cover of thick river fog and smoke, German A7V tanks advanced over no-man's land. An infantry battalion followed up behind the tanks. The Germans crushed British infantry defences destroying machine gun nests and taking numerous prisoners. They broke through a three-mile sector of the Allied lines, captured the town of Villers-Bretonneux, and came to within six or seven miles of Amiens.

As they pushed forward, their tanks encountered three British Mark IV tanks, two of which were females armed only with machine guns and one a male armed with a six-pounder gun and machine guns. "Now all we could see of them, as they lumbered slowly through the fog, was that they were a good deal larger and heavier than the heavy British Tanks, and that they were rather tortoise shaped, the armoured 'shell' everywhere coming down over the tracks like a sort of crinoline" is how Clough Williams-Ellis (1883–1978) described the A7Vs at Villers-Bretonneux in *The Tank Corps*, a book published in 1919 that he co-authored with his wife Amabel (1894–1984).[5] Clough Williams-Ellis, an eminent Welsh architect, served in Belgium and France from 1915 to 1918 in the Welsh Guards and then in the Tank Corps of the British Army.

The leading German tank was named "Nixe" and commanded by Wilhelm Biltz. When the Mark IVs came into view, Biltz ordered Nixe's gunners to fire their cannon and machine guns at the British tanks. The shells damaged the two females but missed the male leaving it unscathed. As the two female tanks retreated, the male Mark IV returned fire with its six-pound guns. Three rounds hit "Nixe" halting it in its tracks, killing one of its gunners, mortally wounding two others of the crew, and inflicting minor wounds on three others.[6] Biltz and the remaining crew abandoned the A7V under machine gun fire from the Mark IV but managed to recover the tank later in the day.

As the Mark IV continued firing on two other A7Vs that had arrived behind "Nixe," seven British tanks known as Whippets joined the attack firing at German machine gun positions and infantry. The Whippets were first produced in 1917 and designed to be lighter and faster than the Mark IVs. Each was equipped with four machine guns.

"The Whippets moved from shell-hole to shell-hole, destroying the machine-gun groups, and then proceeded to deal with the infantry," according to Williams-Ellis.[7] "Their success was terrible. They got right in among the enemy, who had absolutely no cover, and mowed the unhappy Germans down in ranks as they stood. At least 400 of the enemy are estimated to have been killed, and the rest at last fled in confusion, the threatened attack being completely broken up."

German artillery fire eventually disabled the male Mark IV and destroyed three of the Whippets. The fighting around the town continued for another three days until 27 April by which time the Allies had recaptured Villers-Bretonneux and the ground they lost on the first day of the battle.

Wilhelm Biltz survived both the battle and the war. He returned to Clausthal University of Technology after the war and continued his work there until 1921. He then moved to the Technical University of Hanover as head of the inorganic chemistry department. Wilhelm remained single and childless until his death at the age of 66 in Heidelberg on 13 November 1943, just two weeks after his brother died at the age of 78.

REFERENCES

1. W. T. Hall, Ausführung Qualitativer Analysen, By Wilhelm Biltz, (book review), *J. Chem. Educ.*, 1939, **16**, 450.
2. H. Biltz, Über die Konstitution der Einwirkungsprodukte von substituierten Harnstoffen auf Benzil und über einige neue Methoden zur Darstellung der 5.5-Diphenyl-hydantoine, *Ber.*, 1908, **41**, 1379.
3. H. Biltz and W. Biltz, *Laboratory Methods of Inorganic Chemistry*, J. Wiley & Sons, New York, 1909.

4. D. R. Higgins, *Mark IV vs A7V: Villers-Bretonneux 1918*, Osprey, Oxford, 2012, p. 10 and p. 18.
5. C. Williams-Ellis and A. Williams-Ellis, *The Tank Corps*, George H. Doran Company, New York, 1919, p. 262.
6. D. R. Higgins, *Mark IV vs A7V: Villers-Bretonneux 1918*, Osprey, Oxford, 2012, p. 64.
7. C. Williams-Ellis and A. Williams-Ellis, *The Tank Corps*, George H. Doran Company, New York, 1919, p. 263.

CHAPTER 16

Solutions at Sea

ELECTROLYSIS ON BOARD

In 1946, near the end of her life, "the gallant old ship" was still carrying the "original electrolytic tank used in the Dardanelles campaign," and "her medical officer was still proud of the efficacy of the disinfectant solution which it continued to provide." So wrote English biochemist Sir Percival Hartley (1881–1957) in an obituary of English organic chemist Henry Dakin (1880–1952) who had made the original proposal for installing the equipment.[1]

Hartley was commissioned as a lieutenant in the Royal Army Medical Corps (RAMC) in August 1915 and served as a sanitary officer in France. He was promoted to the rank of captain in the spring of 1916. The following year, he saw action at the Battle of Passchendaele that was fought in Belgium from July to November. He was later awarded the Military Cross for gallantry.

His duties on the Western Front included tackling the problems of water supply to troops and dealing with lice infestations. He was also responsible for the disposal of animal carcases, a major problem in the war. On one day alone, 7000 horses were killed during the Battle of Verdun in 1916. Tens of thousands of dogs also died during the war. They were employed as guard dogs on sentry duty, messenger dogs, ratters, scout dogs that detected the scent of the enemy troops, and casualty dogs that

The Chemists' War, 1914–1918
By Michael Freemantle
© Freemantle, 2015
Published by the Royal Society of Chemistry, www.rsc.org

found the wounded on the battlefield and brought them medical supplies.

The Dardanelles campaign to which Hartley referred was a failed attempt by the Allies to force Turkey out of the war. It is also known as the Gallipoli campaign. In February and March 1915, an Allied naval expedition unsuccessfully attempted to pass through the Dardanelle straits in northwest Turkey. The Allies followed up with a military expedition in April 1915, landing troops on the Gallipoli peninsula between the straits and the Aegean Sea. Fighting continued until January 1916 when the Allied troops were withdrawn. Over 50 000 British Empire soldiers and sailors and more than 60 000 Turkish troops died during the campaign.[2]

The "gallant old ship" mentioned by Hartley was the Cunard Line's RMS (Royal Mail Ship) *Aquitania*, a coal-fired ocean liner built at Clydebank, Scotland, and launched on 21 April 1913. It set off on its maiden voyage from Liverpool to New York on 30 May 1914.[3] The ship was requisitioned by the British Admiralty and commissioned into the Royal Navy at the beginning of August 1914. It made three voyages as a troopship to the Dardanelles in the early summer of 1915 before being converted into a hospital ship (Figure 16.1). In July that year, it carried around 2400 wounded soldiers from the Dardanelles to England. The following month the vessel was refitted at Southampton to extend its capacity to over 4000 beds.

The hospital ship was equipped with a special tank for the continuous electrolysis of sea water. The process provided an unlimited supply of hypochlorite disinfectant solution, according to Hartley. The installation resulted in a "great and immediate reduction in the incidence of infections on board," noted Hans Thacher Clarke (1887–1972) in another obituary of Dakin.[4] Clarke was a British-American biochemist who had carried out research on the synthesis of mustard gas when working at the University of Berlin before the First World War (see Chapter 14).

The major components of sea water, apart from water, are sodium and chloride ions. Electrolysis of sea water yields chlorine gas at one electrode, the anode, and hydrogen gas at the other electrode, the cathode, leaving sodium and hydroxide ions in solution. The process therefore converts sodium chloride solution, or brine as it is also called, into sodium hydroxide

Figure 16.1 HMHS (His Majesty's Hospital Ship) Aquitania (The RAMC Muniment Collection in the care of the Wellcome Library).

solution. If the chlorine is allowed to react with the sodium hydroxide at moderate temperatures, a solution containing sodium hypochlorite is formed.

Sodium hypochlorite and the related compound calcium hypochlorite are well known for their bleaching and disinfectant properties. Sodium hypochlorite solution and solid calcium hypochlorite are produced commercially as household bleach and bleaching powder respectively. Chloride of lime, as calcium hypochlorite was known during World War I, was also used extensively as a disinfectant in the squalor of the trenches and in antiseptic preparations to treat the wounded during the war.

Disinfectants and antiseptics are germicides, that is, they kill infectious microorganisms. The term "disinfectant" generally refers to a germicide employed to destroy infectious microbes on or in inanimate surfaces, spaces, objects, or substances such as

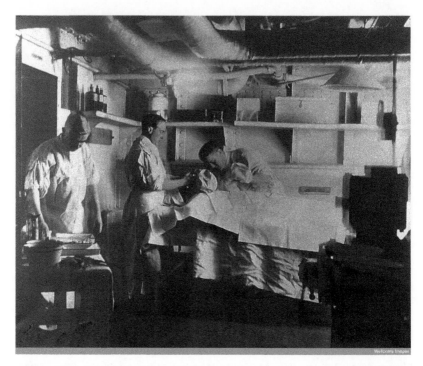

Figure 16.2   Dressing station on HMS Marlborough. A German torpedo hit
the ship on 31 May 1916 during the Battle of Jutland. Two of the
ships's crew were killed and another two wounded (Wellcome
Library, London).

trenches, latrines and dirty water, whereas an "antiseptic" is
applied to skin or wounds to prevent or halt the invasion and
spread of such microbes.

One of the antiseptic solutions widely applied "as a first
dressing in the field" to treat wounds during the First World War
was prepared by dissolving 12.5 grams of bleaching powder and
12.5 grams of boric acid, a weakly antiseptic powder, in a litre of
water.[5] The preparation was developed at the University of
Edinburgh and named "eusol," an acronym for "Edinburgh
University Solution of Lime."

DAKIN'S SOLUTION

The use of disinfectant and antiseptic preparations containing
sodium hypochlorite was pioneered by Dakin although, as Dakin

himself pointed out, the discovery of the chemical dates back to 1788 when French chemist Claude Louis Berthollet (1748–1822) showed that a liquid with bleaching and disinfectant properties could be prepared by the action of chlorine on aqueous alkalis such as sodium carbonate solution.[6]

Dakin was born in London in 1880 and studied chemistry at Yorkshire College, Leeds, which was later to become the University of Leeds. When the war broke out he was carrying out research in a private laboratory in New York. According to Hartley, "he offered his services" for the British war effort but as his health was not good, "the authorities could find no appropriate employment for him." He subsequently joined a group who were carrying out research on the treatment of infected wounds at a temporary military hospital in Compiègne, a town in northern France. The project had been initiated by French surgeon and biologist Alexis Carrel (1873–1944) who won the Nobel Prize in Physiology or Medicine for 1912 for developing new techniques for surgery on blood vessels.

In August 1915, Dakin published his seminal paper "on the use of certain antiseptic substances in the treatment of infected wounds" in which he described the preparation of what became known as "Dakin's solution."[7] The solution was prepared by dissolving sodium carbonate and bleaching powder in tap water, filtering the solution to remove the calcium carbonate that precipitated in the process, and finally adding boric acid. Dakin's solution was not only more stable than other antiseptic preparations used at the time but also less irritant to the skin and wounds.

The solution was used in the so-called Carrel-Dakin treatment of infected wounds. The method involved flushing the whole surface of the wound frequently with a "gentle stream" of Dakin's solution.[8]

According to Hartley, Dakin "was overjoyed to receive the invitation" to help with the British effort in the Dardanelles. In September 1915, Dakin was a member of a medical research committee that travelled to the Gallipoli peninsula on board HMHS (His Majesty's Hospital Ship) *Rewa* to investigate the efficacy of his solution in the treatment of infected wounds.[9] However, the fighting on the peninsula was not intense at the time and so the committee had little opportunity to test the

Figure 16.3   Henry Drysdale Dakin (Wellcome Library, London).

treatment. Most of the patients on board the ship were suffering from dysentery and other diseases. The committee returned to England the following month, but before doing so, Dakin asked the fleet surgeon Frederick Dalton to continue with the investigation.

Over a period of a couple of months, 57 patients with septic wounds on the ship were treated with Dakin's solution that had been prepared on board. No other antiseptic was used. "The usual surgical procedures were adopted for enlarging the wound, making counter openings, removing portions of detached bone, foreign bodies, including bits of clothing, fragments of shell, bullets, gravel, *etc.*" Dalton noted. "Every portion of the wound was then thoroughly irrigated with large quantities of hypo-chlorite solution." The results were "uniformly and consistently good," he added.

Following America's entry into the war in 1917, the United States government requisitioned the *Aquitania* and other British

Figure 16.4   Picture of the sinking of HMHS Rewa published in the 19 January 1918 edition of *The Illustrated London News* (Wellcome Library, London).

ocean liners as troopships to carry troops across the Atlantic to Britain. Whereas the *Aquitania* survived the war, the *Rewa* was not so fortunate. On 4 January 1918, a German U-boat torpedoed and sank the hospital ship in the Bristol Channel (Figure 16.4). The ship was carrying 279 cot and walking wounded cases from Greece to Cardiff. Two of the crew died in the explosion but all the wounded and the remaining crew managed to board lifeboats and were rescued.

REFERENCES

1. P. Hartley, Henry Drysdale Dakin. 1880–1952, *Obit. Not. Fell. R. Soc.*, 1952, **8**, 128.
2. G. Bridger, *The Great War Handbook*, Pen & Sword, Barnsley, 2009, p. 40.
3. Aquitania (1914), 20 March 2014, http://www.maritimequest. com/liners/aquitania_data.htm.

4. H. T. Clarke, Henry Drysdale Dakin. 1880–1952, *J. Chem. Soc.*, 1952, 3319.

5. J. L. Smith, A. M. Drennan, T. Rettie and W. Campbell, Antiseptic action of hypochlorous acid and its application to wound treatment, *Br. Med. J.*, 1915, **2**, 129.

6. H. D. Dakin, The antiseptic action of hypochlorites, *Br. Med. J.*, 1915, **2**, 809.

7. H. D. Dakin, On the use of certain antiseptic substances in the treatment of infected wounds, *Br. Med J.*, 1915, **2**, 318.

8. Anon., The war: Carrel-Dakin treatment of wounds, *Br. Med. J.*, 1917, **2**, 597.

9. F. J. A. Dalton, Sodium hypochlorite in the treatment of septic wounds, *Br. Med. J.*, 1916, **1**, 126.

CHAPTER 17

America's Wartime Potash Problem

AN ABUNDANCE OF POTASH

Although Germany experienced shortages of the fixed nitrogen it required for the manufacture of explosives and food production during the First World War, it had an abundance of a naturally occurring material containing another element of critical importance, especially in agriculture. That was potassium, an element that occurs in a variety of minerals collectively known as potash.

Potash is defined in various ways and its composition is normally but curiously given in terms of its potassium oxide (K_2O) content. One science and technology dictionary on my shelf defines potash as: "Potassium, usually expressed notionally as K_2O, in a fertilizer."[1] Another, a chemical dictionary, has the following entry for potash: "(1) Potassium hydroxide. (2) Potassium carbonate. (3) An early name (pot ash) for wood ash used as a source of potassium carbonate."[2] And yet another dictionary of chemistry, strangely in view of the importance of potash, has no entry for it at all.[3]

Potassium, like nitrogen and phosphorus, is a primary macronutrient that plants absorb from the soil in relatively large amounts. The element helps plants to grow, regulate their use of

The Chemists' War, 1914–1918
By Michael Freemantle
© Freemantle, 2015
Published by the Royal Society of Chemistry, www.rsc.org

water, and resist diseases and marauding insects. Water-soluble potassium salts, began to be widely used as soil fertilisers towards the end of the 19th century. In 1865, German chemist Justus von Liebig (1803–1873), who is considered to be the "father of the fertiliser industry," had stressed the importance of these compounds as fertilisers and advocated their increasing application in agriculture. The most widely used potassium fertilisers were and still are potassium chloride (also known as muriate of potash) and potassium sulfate (sulfate of potash). They contain the equivalent of 60% and 50% K_2O, respectively.

Figure 17.1 Justus von Liebig, "father of the fertiliser industry" (Royal Society of Chemistry Library).

Whereas nowadays potash is extracted from deposits throughout the world, until the beginning of the 20th century, Germany was the only country to practise potash mining. In 1843, brine (salty water) containing the salts potassium chloride and magnesium chloride was discovered at Stassfurt, a small German town near Madgeburg, where common salt (sodium chloride) was already being mined.[4] In the 1850s potassium salt rocks containing various minerals were found in the so-called Magdeburg-Halberstadt rock salt basin. They were called Abraum salts. One of the most important of these salts was carnallite, a mineral composed of magnesium chloride, potassium chloride, and water. Another was kainite, which consisted of magnesium sulfate, potassium chloride, and water.

The first factory to produce potash from these deposits was erected in Stassfurt in 1861. It marked the birth of potash mining. The following year, the output of potash salts was just under 20 000 tons. By 1909, production had soared to over 7 million tons.[5] In the year before the outbreak of the First World War, Germany employed some 30 000 workers in its potash mines.[6]

Crude potash salts were principally sold as fertilisers but large amounts were also refined in so-called "potassium chloride factories" to produce a range of potassium salts. Crystalline potassium chloride, for example, was readily produced by dissolving the crude carnellite mineral in water and carrying out a sequence of separation and crystallisation processes.

Potash was a commercially important raw material not just in agriculture but also in many other industries. "In the many-sided technical and industrial life of today there are but few enterprises that can dispense with the products of the potash industry," commented W. C. Ebaugh, in a paper on potash presented to the Utah Society of Engineers, Salt Lake City, on 18 April 1917.[7] He pointed out that it was employed in medicine, photography, painting, dyeing, laundry work, bleaching, spinning works, soap manufacture, refrigeration, preservatives, electrotechnics, fireworks, explosives, matches, paper, glass, aniline colours, and metallurgy.

It was an impressive list. The use of potash to make soap is a classic example of its economic value. Soap is produced by boiling a fat or an oil with an aqueous solution of an alkali. When sodium hydroxide (also known as caustic soda) is used as the alkali, a hard soap is produced. If the alkali is potassium hydroxide (caustic potash), a soft or liquid soap is produced. Potassium hydroxide was traditionally prepared by the reaction of potassium carbonate with calcium hydroxide (slaked lime). Alternatively, potassium hydroxide can be prepared by the electrolysis of an aqueous solution of potassium chloride.

The heads of safety matches, widely used since the 19th century, typically contain about 50% potassium chlorate, a compound derived from potassium chloride. Potassium dichromate, a bright orange-red compound that can be prepared from either potassium chloride or potassium hydroxide, was used for tanning leather, dyeing textiles, and in photography and a range of other applications.

THE POTASH CRISIS

"The German mines, located principally in the neighbourhood of Stassfurt and Magdeburg in Prussian Saxony have been for the past seventy years the principal source of the world production of the various potassium salts," noted A. S. Cushman and G. W. Coggeshall in a paper presented at the 7th annual meeting of the American Institute of Chemical Engineers, held in Philadelphia in December 1914.[8] The authors noted that the United States imported millions of dollars' worth of potash salts such as potassium chloride, potassium sulfate, and potassium carbonate as well as potassium hydroxide. About 85% of the imported potassium chloride was used in the fertiliser industry, the remainder being used for other industrial purposes. For example, about 5% of the imported potassium chloride was converted into potassium dichromate.

The United States was by far the largest consumer of exported German potash. It imported potash equivalent to more than 250 000 tons of K_2O in 1913.[9] The potash was purchased from the German Potash Syndicate. The syndicate, or Kartell as it was also known, was initially established in 1888 to fix maximum domestic prices for potash in Germany and set minimum export prices.

Germany's world monopoly of the potash market created a crisis in the United States after the outbreak of war when the German government imposed an embargo on potash exports and Britain enforced a naval blockade of German ports. Within a few months, the price of potash in the United States rocketed from just over $11 per ton to $100 per ton.

In some cases, American potash users were able to side-step the problem by replacing potassium compounds with sodium compounds. The chemical properties of potassium hydroxide and sodium hydroxide are very similar, for example. The same goes for potassium permanganate and sodium permanganate, both of which are powerful oxidants. Gas masks developed by the American Chemical Warfare Service in 1917 and 1918 incorporated canisters containing sodium permanganate, rather than potassium permanganate, and soda lime (a mixture of calcium hydroxide and sodium hydroxide). The permanganate oxidant destroyed toxic organic gases while soda lime, a strongly alkaline material, neutralised acidic gases such as chlorine. By the end of the war, the United States had manufactured over five million gas masks with canisters containing these chemicals.

Using sodium compounds instead of potassium compounds may work in laboratories or in gas masks, but it does not work in the soil. The total cessation of shipments of potash from Germany "seriously disabled the American fertilizer industry" and "resulted in material injury to the agricultural sections using fertilizers, particularly the cotton belt in the South," reported Raymond F. Bacon in a paper on "The war and American chemical industry" presented to the Urbana Meeting of the American Chemical Society (ACS) on 18 April 1916.[10]

THE SEARCH FOR OTHER SOURCES

The price and scarcity of potash drove consumers, notably farmers such as the cotton planters of the Southern states, to look for other sources of supply. The crisis also sparked intense research that led to the launch of a variety of private industries to meet the demand for potassium salts.

Potash was known to occur in the materials used to make cement and also as a by-product in the production of molasses and beet sugar. It was found in seaweed in the coastal waters of California and in a number of American desert lakes. Banana stalks were identified as another possible commercial source of potash. The stalks were classified as a waste material but their ash "contained no less than 45.9% potash" noted one report on the research published in 1916.[11]

The files of the United States Patent Office "show that the question of supplying our own potash has greatly exercised the brains of inventors," observed Edward Hart in 1915.[12] The exercise continued with increasing vigour throughout the war.

"Chemistry was called upon to help out in this real emergency and nearly every possible source of potash, both organic and inorganic, was given consideration and close study," remarked William H. Waggaman, a fertiliser scientist at the Bureau of Soils in Washington, DC.[13] "The feverish energy expended in research was not long in bearing fruit."

The year 1915 marked the beginning of the American potash industry, A. W. Stockett noted in a paper summarising developments in the industry that he presented at a symposium on potash held by the Division of Industrial Chemists and Chemical Engineers at the 58th ACS national meeting, held in Cleveland, Ohio, on 12 September 1918.[14] He reported that the industry produced just over 1000 tons of K_2O in 1915. By 1917, the total had risen to over 32 000 tons.

Much of the potash was extracted from potash-rich brine lakes in the Sandhills region of Nebraska. Before the war, the region had been sparsely populated but by 1917 ten potash plants had been set up in the region.[15] Production reached a peak during the year with potash selling for more than $150 per ton. Antioch, a town near several of the lakes, turned into a boomtown with its population increasing from 175 to several thousand within three years. It became known as the capital of the potash industry in Nebraska.

Each plant had a capacity to produce 100 tons of potash per day. The extraction process was fairly simple. Brine was pumped from the lakes through pipes to the plants where it was concentrated by solar evaporation and boiling. The concentrated liquor was then cooled allowing the potash to crystallise out. The wet crystalline potash was finally dried. Most of the material was used as fertilizer in the American cotton belt in the South.

A similar process was employed by the American Trona Corporation which, in 1916, began producing potash from mineral deposits in Searles Lake in California. The deposits also contain an abundance of soda ash (sodium carbonate) minerals and borax, a widely-used mineral containing the element boron. The corporation produced some 1800 tons of crude potash salts per month, according to one of its workers, Alfred de Ropp, Jr, who presented a paper on the topic at a symposium on potash held at the 4th National Exposition of Chemical Industries in New York, 25 September 1918.[16]

J. W. Hornsey, another contributor to the symposium, noted that other American desert lakes were possible sources of potash. They included Owens Lake in California, Great Salt Lake in Utah, and Abert Lake and Summer Lake in Oregon.[17] He pointed out, however, that the brines from these lakes contained contaminating salts that made commercial potash production difficult.

Hornsey's presentation also outlined the use of alunite as a source of potash. Alunite is a sulfate mineral rich in potassium and aluminium. In 1915, the Minerals Products Corporation built a plant near the town of Marysvale, Utah, to exploit the alunite deposits there. The small town that sprang up around the plant was named Alunite after the mineral.

Potash was extracted from alunite by crushing the mineral, roasting it in a kiln, and then soaking the resulting mixture of potassium sulfate and alumina (aluminium oxide) in a tank of hot water. Potassium sulfate dissolved in the water leaving behind the insoluble alumina which was separated by filtration. Evaporation of the solution yielded crystalline potassium sulfate which was then dried in a centrifuge. According to Stockett, a little over 2400 tons of K_2O in the form of high-grade sulfate were produced from alunite between 1915 and 1918.

At the same symposium, C. A. Higgins of the Hercules Powder Company described the recovery of potash from kelp.[18] The vast forests of large seaweeds that grew in the Pacific Ocean off the California shore offered "inexhaustible supplies of potash" he said. He outlined two methods employed to extract potash from the kelp. The first involved drying and incinerating the seaweed to produce ash containing around 15% K_2O. The process proved expensive, however. In 1915, the company constructed a factory at San Diego designed to yield potash far more economically. The process relied on fermentation of the kelp to produce a liquor containing not only potash but also other compounds that could be readily converted into commercially valuable chemical products such as acetone and iodine. According to Stockett, giant kelps harvested from the Pacific coastal waters accounted for 11% of the total American potash production in 1917.

Potassium feldspar, a potash-rich silicate mineral that also contains aluminium, was investigated as a potential source of potash. Although numerous patents were issued for extracting potash from the mineral, no commercially-viable process was developed.

The dust from blast furnaces was considered to be another promising source of potash. The iron ore, coke, and limestone loaded into the furnaces all contained small amounts of potash, typically less than 1% in terms of K_2O. Even so, with the enormous amounts of these materials used for iron and steel production during the war, an estimated 200 000 to 300 000 tons of K_2O per year was potentially available from this source, Stockett noted. The industry, however, was not interested. "All of the manufacturers are at present so intent on producing the maximum amount of pig iron that there is very little possibility of getting them to realize the importance of developing a domestic potash industry," commented Stockett at the potash symposium.

The dust from cement kilns proved to be a more attractive proposition. An investigation by the Bureau of Soils estimated that up to 100 000 tons of K_2O might be recovered from the country's cement works. The Riverside Portland Cement Company of California was the first company to recover potash from kiln dust and it did so almost serendipitously.

Before the war, orange growers near the company's cements works in Los Angeles claimed that fine dust escaping through the kiln flues was damaging their fruit trees. Following litigation, the company was required to take measures to trap the flue dust and in 1913 installed a Cottrell electrostatic precipitation plant to collect the dust.

The precipitation process, which is still used today to control industrial particulate emissions, involves passing dust particles through an electric field. The particles become electrically charged and are attracted to and collected at plate electrodes with the opposite electrical charge. The electrostatic precipitator was invented by American chemist Frederick Gardner Cottrell (1877–1948) and described in a patent with the title "Art of separating suspended particles from gaseous bodies" on 11 August 1908.[19]

To the cement company's surprise, analysis of dust samples collected from the kiln flues revealed a significant potash content, up to 16% in some cases. The potash originated from the various clay minerals in the limestone used to make the cement. The value of the trapped dust soon became widely recognised and by the end of the war some dozen American cement companies had installed Cottrell dust recovery plants at their works. They yielded a total of between 10 000 and 12 000 tons of K_2O each year, according to Stockett.

COLLAPSE OF THE POTASH INDUSTRY

Total annual potash production in the United States reached a peak of 54 000 tons of K_2O in 1918, but with the end of the war, the price dropped to a fraction of its price as consumers were reluctant to buy American potash. Many opted to wait for supplies of imported German potash which they predicted would be available at much cheaper prices. Plants were forced to close down in 1919 and production in the United States for the year dropped to 32 000 tons of K_2O, but imports did not arrive and so some of the plants reopened. Production increased to 48 000 tons of K_2O in 1920 but then collapsed dramatically to 10 000 tons in 1921.[20]

Fertiliser manufacturers once again stopped buying American potash in late 1920 and 1921 as they became more optimistic that they would be able to purchase the product at lower prices from Germany. At the time Congress was debating a tariff bill. The fertiliser manufacturers who purchased potash lobbied furiously against a protective tariff on imported German potash. Potash producers on the other hand lobbied hard to have imported potash listed in the bill. A tariff would enable them to stay in business.

"Of the 128 plants reporting production in 1918, over 100 made nothing in 1921," reported John E. Teeple (1874–1931), an American chemist and potash expert.[21] Some of the plants were "abandoned as war babies that could not live in peace times," he commented. While many companies ended in bankruptcy, a few kept going in the hope that Congress would protect potash production. It did not. In September 1922 the United States government passed its tariff bill. Potash was not listed in the bill. In 1924, only eleven plants were producing and selling potash in the United States, their total production being just short of 23 000 tons of K_2O.

The collapse of the potash industry had other consequences as well. Potash boom towns such as Antioch and Alunite were largely abandoned as plants closed down. They became ghost towns.

REFERENCES

1. *Larousse Dictionary of Science and Technology*, ed. P. M. B. Walker, Larousse, Edinburgh and New York, 1995.
2. *Grant & Hackh's Chemical Dictionary*, ed. R. Grant and C. Grant, 5th edn, McGraw Hill, New York, 1987.
3. D. B. Hibbert and A. M. James, *Macmillan Dictionary of Chemistry*, Macmillan, London and Basingstoke, 1987.
4. J. R. Partington, *The Alkali Industry*, 2nd edn, Van Nostrand, New York, 1925, p. 309.
5. H. R. Tosdal, The Kartell movement in the German potash industry, *The Quarterly Journal of Economics*, 1913, **28**, 140.
6. F. Aftalion, *A History of the International Chemical Industry; from the "Early Days" to 2000*, Chemical Heritage Press, Philadelphia, 2001, p. 103.
7. W. C. Ebaugh, Potash and a world emergency, *Ind. Eng. Chem.*, 1917, **9**, 688.
8. A. S. Cushman and G. W. Coggeshall, Feldspar as a possible source of American potash, *Ind. Eng. Chem.*, 1915, 7, 145.
9. W. H. Bowker, The agriculture industries, *Ind. Eng. Chem.*, 1915, 7, 59.
10. R. F. Bacon, The war and the American chemical industry, *Ind. Eng. Chem.*, 1916, **8**, 547.
11. Anon., New source of potash, *Ind. Eng. Chem.*, 1916, **8**, 743.
12. E. Hart, The potash situation, *Ind. Eng. Chem.*, 1915, 7, 670.
13. W. H. Waggaman, The fertilizer industry, *Ind. Eng. Chem.*, 1922, **14**, 789.
14. A. W. Stockett, The potash situation, *Ind. Eng. Chem.*, 1918, **10**, 918.
15. R. E. Jensen, Nebraska's World War I potash industry, *Nebraska History*, 1987, **68**, 28.
16. A. de Ropp, Jr., Potash from Searles Lake, *Ind. Eng. Chem.*, 1918, **10**, 839.

17. J. W. Hornsey, Potash from desert lakes and alunite, *Ind. Eng. Chem.*, 1918, **10**, 838.
18. C. A. Higgins, Recovery of potash from kelp, *Ind. Eng. Chem.*, 1918, **10**, 832.
19. F. G. Cottrell, *US Pat.*, 895 729, 1908.
20. N. Sheeve, Potash, *J. Chem. Educ.*, 1927, **4**, 230.
21. J. E. Teeple, Potash in America – two years' progress, *Ind. Eng. Chem.*, 1922, **14**, 787.

CHAPTER 18

Fractured Friendships

NO QUARREL WITH GERMANY

On the afternoon of Sunday 2 August 1914, two days before Britain declared war on Germany, a peace rally took place in Trafalgar Square, London. It was the largest rally that had been held there in years and was one of several "Stop the War" rallies in towns and cities around the country at the beginning of August. Scottish Labour politician Keir Hardie (1856–1915), a pacifist and Member of Parliament for Merthyr Tydfil, Wales, spoke at the Trafalgar Square rally. The crowd cheered loudly when he called for a general strike if Britain declared war. "You have no quarrel with Germany," he roared.

Most chemists in Britain before the war would have agreed with Hardie. Not only did British chemists have no quarrel with Germany, many travelled there to do post-graduate research. When English chemist Arthur Smithells (1860–1939) graduated, for example, he went to work in the laboratory of Adolf von Baeyer, a chemistry professor at the University of Munich. Smithells had studied chemistry at Owens College, Manchester, and received a Bachelor of Science degree from the University of London in 1881. "It was the custom in those days to obtain post-graduate training in Germany, so after graduation Smithells along with two of his fellow-students . . . went to Munich to study

The Chemists' War, 1914–1918
By Michael Freemantle
© Freemantle, 2015
Published by the Royal Society of Chemistry, www.rsc.org

under von Baeyer," commented his Royal Society obituarist Henry Raper.[1]

Baeyer was highly respected in Britain and famed throughout the world of chemistry for his research on indigo and his syntheses of the blue dye by two different routes in 1878 and 1880. In the 1880s he was honoured by the American Academy of Arts and Sciences and in Britain by the Royal Society of London. He subsequently won the Nobel Prize in Chemistry for 1905 for his work on synthetic organic dyes and his contributions to the "advancement of organic chemistry and the chemical industry." When young, Baeyer had studied chemistry at the University of Heidelberg under Robert Bunsen who invented the Bunsen burner.

After a few months working with Baeyer, Smithells went on to carry out further research in Bunsen's laboratory in Heidelberg. In 1883, Smithells returned to Manchester to take up an appointment as assistant lecturer in chemistry at Owens College

Figure 18.1   Adolf von Baeyer (Royal Society of Chemistry Library).

and two years later was appointed professor of chemistry at Yorkshire College, Leeds, which was to become the University of Leeds in 1904. His research initially focused mainly on the processes of combustion, particularly when hydrocarbon gases such as methane and ethene were burnt in a Bunsen burner. He was elected a Fellow of the Royal Society in 1901.

"The outbreak of war in 1914 was a great shock to Smithells, as it was to many others who had been trained in Germany and had long-standing friendships with her men of science," observed Raper who had studied chemistry at Yorkshire College and later became lecturer in physiological chemistry at the University of Leeds.

Like Smithells, English chemist Arthur Crossley also studied chemistry at Owens College, Manchester. He graduated there with an honours degree in 1890 and the following year went to

Figure 18.2   Arthur Smithells (Royal Society of Chemistry Library).

Germany to carry out research at the University of Würzburg with future German Nobel Chemistry Prize winner Emil Fischer who had been one of Baeyer's doctoral students in the 1870s. Crossley obtained a PhD degree at the university and then moved with Fischer to the University of Berlin in 1892. In early 1893, Crossley returned to Owens College and worked with William Henry Perkin, Jr, the son of Wiliam Henry Perkin, Snr, who had synthesised mauve, the first synthetic coal-tar dye, in 1856. In June 1914, Crossley was appointed professor of organic chemistry at King's College, London.

James McConnan (1881–1916), another English chemist, also obtained a PhD degree in Germany. He studied chemistry at University College, Liverpool. After graduation in 1902, he proceeded to the University of Jena where he carried out his doctoral research for two years under the supervision of German chemistry professor Ludwig Knorr. McConnan then returned to England as assistant lecturer and demonstrator in organic chemistry at the University of Liverpool which had become an independent university the previous year. In 1907, McConnan took up a post as research chemist for the River Plate Fresh Meat Company in Argentina. When his father died in 1913, James returned to England to succeed him as director of the rope and twine manufacturer, Jackson, McConnan and Temple. At the same time, he resumed his research activities in organic chemistry at the University of Liverpool and then, before the outbreak of war, joined the research staff of the soap makers Lever Brothers at Port Sunlight in northwest England.

Smithells, Crossley, and McConnan were just three of numerous British chemistry graduates who went to Germany for training with some of Germany's most eminent chemists in the late 19th and early 20th centuries. Germany also attracted young up-and-coming chemists from other countries, not least the United States.

Russian-born American chemist Moses Gomberg (1866–1947), for example, took a year's leave of absence as an instructor at the University of Michigan, Ann Arbor, to carry out postdoctoral research in Germany in 1896 and 1897.

Gomberg was born into a Jewish family in Elisavetgrad, a city in central Ukraine now known as Kirovohrad. After the assassination of Russian emperor Alexander II in 1881, the government suspected Gomberg's father, George, of political conspiracy. In 1884, the family farm was confiscated and Moses, now aged 18, also came under suspicion. To avoid arrest and imprisonment, George fled to the United States and settled in Chicago, Illinois. Moses accompanied his father or followed him soon after—the details are not clear as little is known of his early life.[2] The young Gomberg graduated with a Bachelor of Science degree from the University of Michigan in 1890 and earned his PhD there in 1894.

In Germany, Gomberg worked for two semesters in Baeyer's laboratory in Munich and then a further semester with Viktor Meyer, professor of chemistry at the University of Heidelberg. Meyer had synthesised mustard gas some ten years earlier while at the University of Göttingen (see Chapter 14). Gomberg returned to Michigan after his stay in Germany and was appointed assistant professor and then full professor at the University of Michigan in 1899 and 1904, respectively. He is most famous for the discovery of triphenylmethyl, a stable organic free radical, in 1900. A free radical, sometimes known simply as a radical, is an atom or molecule with an electron that has not paired up with another electron to form a chemical bond.

The American Chemical Society (ACS) designated Gomberg's discovery as a National Historic Chemical Landmark at a ceremony held at the University of Michigan in June 2000.[3] Gomberg "challenged the then prevailing belief that carbon could have only four chemical bonds," notes a plaque commemorating the event. "Today, organic free radicals are widely used in plastics and rubber manufacture, as well as medicine, agriculture and biochemistry." Gomberg went on to become ACS president in 1931.

NO QUARREL WITH BRITAIN

Paradoxically, the rapid growth of German chemical industry in the second half of the 19th century and early 20th century combined with the development of a large number of well-equipped academic chemistry laboratories and the high reputation of science education in the country not only encouraged young British and American chemists to spend a year or two doing research in Germany, it also encouraged German chemists to find jobs in other countries. The scientific infrastructure in Germany, the new prestige of being a chemist, and the opportunities in the chemical industry led in the 1850s to a "disproportionate number of graduates" that exceeded demand, according to German chemist Walter Wetzel.[4] The surplus was "forced to look abroad for employment and the nearest possibility" was the United Kingdom.

Wetzel has produced evidence revealing that "a large number of German chemists not exactly determinable" were employed in British establishments from 1845 until 1914.[5] The establishments included alkali works, breweries, dyestuff companies, sugar refineries, copper mills, museums, private laboratories and colleges such as Owens College, Manchester.

One of the most influential German chemists to work in England during this period was August Wilhelm von Hofmann. In 1845, he became the first director of the newly-founded Royal College of Chemistry in London. Hofmann was the "father of a generation of British researchers," notes Wetzel. Perkin, for instance, who discovered mauve in 1856, studied and worked under Hofmann. The German chemist was "one of the giants of 19th century organic chemistry" who "also made major contributions to chemical education," according to the authors of a centennial tribute to him published in 1992.[6]

The following lengthy quote from the tribute illustrates the extent of his contributions and diverse nature and scope of applied chemistry at the time: "Hofmann's chemistry syllabus at the college for 1853–56 focused on the textile industry and the applications of organic chemistry, covering: wood, peat, starch, sugar, and gums; fermented liquors; organic colouring matters; bleaching, dyeing and printing of textile materials and leather; oils, vegetable and animal, and their products, such as candles and soap; ammoniacal salts; prussiate of potash and Prussian blue; gelatine; animal black *etc.*"

And the German contribution to education and training in science in Britain did not stop there. As late as April 1913, an unnamed German scientist lectured staff at a naval torpedo and mining establishment at Portsmouth "on the merits of TNT as a filling for shell, mines or torpedoes," notes British author Guy Hartcup in his book on scientific developments during the First World War.[7]

FRIENDS BECOME FOES

When war broke out, the exchange of chemists between Germany and Britain and other Allied countries came to an abrupt halt. Chemists who had once been friends and colleagues now became foes. Some contributed their chemical expertise to their respective country's war efforts while others took up arms to fight.

When the Germans mounted their first gas attack at Langemarck in April 1915, Smithells offered his services to Northern Command, a Home Command of the British Army. Northern Command assigned him the task of visiting troops in its camps in northern England and teaching them about chemical warfare. Smithells was subsequently appointed chief chemical adviser for anti-gas training of the Home Forces and transferred to the army's general headquarters based at the War Office in London.[8] In 1916 he attained the rank of lieutenant-colonel in the army and in 1918 was awarded the CMG (Companion of the Order of St Michael and St George) for his services during the war.

Raper, Smithells' Royal Society obituarist, enlisted in the Royal Army Medical Corps (RAMC) and later transferred to the Royal Engineers where he worked on anti-gas measures as a commanding officer.[9] He achieved the rank of lieutenant-colonel and was rewarded with a CBE (Commander of the Most Excellent Order of the British Empire) for his services.

Crossley also became involved in chemical warfare. In the early months of the war, he initially acted as secretary to scientific and industrial committees that advised the Ministry of Munitions on the production and supply of chemical materials for the British war machine. Then, in November 1915, he was appointed liaison officer for chemical warfare with the rank of lieutenant-colonel and travelled on several fact-finding missions to the Western Front battlefields where chemical weapons had been used.

In June the following year, Crossley was appointed head of "The War Department Experimental Ground," an experimental chemical warfare station that the Ministry of Munitions had established in March that year. He was given the task of converting 3000 acres of bare land at Porton Down in the middle of Salisbury Plain in Wiltshire into an experimental ground, staffing and equipping the site, and supervising the experiments there. On arrival, Crossley discovered that the establishment consisted of just two small army huts with no water, no equipment whatsoever, and no roads leading to them.[10] Within weeks he had set up a chemical laboratory in one of the huts and by the end of the year a Royal Engineers experimental company, assisted by a Royal Artillery Experimental Battery and RAMC officers, began to carry out testing at the site.[11]

Over the next two years, Crossley's team established chemical, anti-gas, and physiological laboratories at Porton, as well as animal houses, a field trials department, a meteorological section, workshops, gun sheds, magazines, barracks and canteens. Permanent roads to access the site were built and telephones, electricity and water laid on. By the end of the war the establishment had expanded to a large hutted camp on 6000 acres with a military staff of around 1000 officers and other ranks and a civilian staff of 800. Crossley and his group tested around 200 chemical warfare agents for their persistency, toxicity and other properties and also developed gas masks that provided protection against new poison gases that the Germans were introducing on the Western Front. They not only carried out thousands of experiments but also test-fired some 40 000 rounds of ammunition.

Whereas British chemists Smithells, Raper and Crossley all survived the First World War, McConnan, like so many other British chemists at the time, was not so fortunate. At the outbreak of war, McConnan, "with instinctive patriotism," enlisted as a private in the British Army and after a period of training was rapidly promoted to the rank of sergeant.[12] In March 1915, he received a commission as an officer with the rank of second-lieutenant and six months later was sent to fight in the Gallipoli campaign against the Turkish Army. In July the following year, he went with his battalion to the Western Front where, "owing to his excellent knowledge of French," he served for a short time as a billeting officer. Having just been promoted to the rank of temporary captain, he was killed in France on 9 September 1916 during the Battle of the Somme while organising a patrol to reconnoitre a farm. His name appears on the Chemical Society war memorial in the Royal Society of Chemistry's headquarters in London (see Chapter 19).

Gomberg became a major in the Ordnance Department of the United States Army during the war.[2] He served as an adviser on the production of propellants and high explosives and also devised a method for the manufacture of 2-chloroethanol, also known as ethylene chlorohydrin, a chemical used to make mustard gas by the Meyer–Clarke route (see Chapter 14).

THE MANIFESTO OF THE NINETY-THREE

The close friendship between German chemist Emil Fischer and British chemist William Ramsay, winners of the Nobel Prize in Chemistry in 1902 and 1904, respectively, became one of the most notable casualties of the First World War. Four years after graduating at the University of Glasgow at the remarkably young age of fourteen, Ramsay travelled to Germany where he studied chemistry under German organic chemist Wilhelm Rudolph Fittig (1835–1910) at the University of Tübingen. He obtained his doctorate at the university in August 1872—at the age of nineteen—and soon developed a great affection for Germany, admiring its science, education system, and culture. Fluent in both spoken and written German, he paid a number of visits to the country before the war and became acquainted with a number of renowned German chemists. His son studied in Germany and his daughter and son-in-law lived in the country for a while.

Fischer similarly had the greatest respect for Britain and British chemists. He regularly corresponded with many British chemists and visited Britain several times. His oldest son Hermann studied at Cambridge University. Fischer and Ramsay, who were both the same age, first met in December 1898 and got on well. Thereafter they met at scientific meetings whenever they could, wrote to each other, and their families visited each other's homes.[13]

With the outbreak of war, Fischer and Ramsay both became embroiled in the patriotic fervour of their respective countries. As a consequence, their friendship tore apart acrimoniously. Fischer was one of the signatories of the notorious German "Manifesto of the Ninety-Three." The proclamation, dated 23 October 1914, was signed by 93 eminent representatives of the German sciences and arts. It protested against the "lies and calumnies" of the Allies who were "endeavouring to stain the

Figure 18.3   Emil Fischer (Royal Society of Chemistry Library).

honour of Germany in her hard struggle for existence—a struggle that had been forced upon her." The manifesto claimed that Germany was not guilty of causing the war; justified Germany's march through Belgium; asserted that the life and property of not one single Belgian citizen was injured or destroyed by its soldiers; and condemned those who allied themselves with Russians and Serbians and incited "Mongolians and negroes against the white race." Haber, Baeyer, Ostwald, and Willstätter were among the other German Nobel Prize winners (see Chapter 4) who signed the manifesto.

Ramsay and other British scientists were outraged and responded to the German proclamation with their own countermanifesto entitled: "Reply to German professors."[14] It blamed Germany and its "unfortunate ally" Austria-Hungary for deliberately provoking the war, violating Belgian neutrality, and destroying or bombarding monuments of human culture. Britain, Belgium and its allies, it asserted, were waging the war for defence, liberty and peace. A total of 117 British intellectuals, including Ramsay and Perkin, Jr, signed the document.

Figure 18.4   William Ramsay (Royal Society of Chemistry Library).

The German manifesto also prompted Ramsay, who had been such a Germanophile before the war, to make vehement Germanophobic public statements, notably in a series of letters to the British national newspaper *The Times* published in 1914 and 1915. His letter to the newspaper on 24 October 1914 concerning "The 'Zeppelin' scare," was typical.[15] Zeppelins were German airships built with fabric-covered rigid aluminium alloy frameworks containing cells filled with hydrogen. The first Zeppelin bombing mission against the Allies was carried out in August 1914 when the crew dropped artillery shells on Liege in Belgium. Zeppelins subsequently bombed a number of towns and villages on the east coast of England in the early months of 1915.

In his letter of October 1914, Ramsay, the discoverer of inert gases such as helium, first discussed the use of hydrogen and the possible use of other gases to fill the airships. He then moved on to the nature of the metal frameworks. Finally, he angrily called for any Zeppelin crews that were captured to be hanged. "That there is a small danger from bombs dropped from Zeppelins cannot be denied," he wrote. "But it would be well that short shrift should be given to any of the crew of a Zeppelin which might be captured after dropping bombs. This is not war; it is murder; and the statutory penalty is death by hanging. If it were generally known that such a penalty were the invariable consequence of being one of a Zeppelin's crew, it might deter men from volunteering on such barbarous adventures."

Figure 18.5   St Bartholomew the Great church damaged by a 660-pound bomb dropped from a Zeppelin during an air raid on 8 September 1915. The church in West Smithfield dates back to 1123 and is London's oldest parish church (Wellcome Library, London).

Ramsay continued to attack the Germans in further vitriolic letters to *The Times* and in articles in other publications published in 1915 and early 1916. He died of cancer in July 1916.

Following publication of the Manifesto of the Ninety-Three, Fischer began to have misgivings about the document and by the beginning of 1915 he started to call for a resumption of international scientific cooperation when the war was over. As the war progressed, however, Fischer increasingly suffered from poor health, insomnia and depression. Some of it was brought on by the toxic chemicals with which he worked. Furthermore, two of his three sons died during the First World War, one from suicide. On 15 June 1919, the Nobel laureate took his own life by poisoning himself with hydrogen cyanide.

SUSPICION AND HYSTERIA

In Britain, the start of the First World War unleashed widespread public feelings of not only hostility towards the enemy but also of suspicion of German nationals residing in the country. On 22 August 1914, for instance, *The Times* published a letter by a person named "S" about "Highly-placed spies" in England.[16] "S" suggested that "during the last quarter of a century numbers of highly-placed aliens, some naturalized, some not, who are known to be in close communication with German financial and political circles, have bought their way into English Society." The writer noted that it was quite possible to communicate with Berlin by cable *via* the United States and advised "the authorities carefully to watch the correspondence, telephonic communications, telegrams, and cablegrams that may be emitted directly or indirectly by these highly-placed German sympathizers."

The anti-German feelings in Britain intensified to a level of hysteria following the sinking of RMS (Royal Mail Ship) *Lusitania* off the south coast of Ireland by a German U-boat on 7 May 1915 with the loss of 1198 lives. The predicament of a German chemist employed at Cambridge University at the time is just one example of the impact of this hysteria. "Many people are asking what the authorities intend to do about German members of the University—graduate and undergraduate," reported the *Cambridge Daily News* on 15 May. "There is at least one German drawing a good salary from the University who has never made any pretence of concealing his anti-British sympathies." The report added that the university would probably receive a government subsidy towards chemical research work. "It is to be hoped that before parting with the taxpayers' money for this purpose the Government will see to it that the proposed researches are carried out by Britons, and that the results will not be accessible to any German employed in the Chemical Laboratories." The quotes are taken from a passage of the 15 May news report reproduced in an article about Ruhemann published in 1993.[17]

There is little doubt that the German mentioned in the report was Siegfried Ruhemann (1859–1943), a lecturer in organic chemistry at Cambridge University. Ruhemann was born in the East Prussian town of Johannisburg (now Pisz, Poland). He graduated in 1882 and obtained his PhD under Hofmann at the University of Berlin in 1881. He continued working with Hofmann until 1885 when he moved to Cambridge after

Figure 18.6   Siegfried Ruhemann (Royal Society of Chemistry Library).

Hofmann had recommended him for a position as assistant to Scottish chemist and physicist James Dewar, a professor at the university.

In 1911, Ruhemann made his most famous discovery—the observation that the organic compound ninhydrin reacted with amino acids and peptides to give a purple dye, later known as Ruhemann's purple, which could be used as a reagent to characterise these compounds.[18] He married Olga Liebermann in Berlin in 1900 and became a naturalised British citizen in 1903 following the birth of his son, Martin. On 7 May 1914 he was elected a Fellow of the Royal Society of London.

The war broke out three months later. Ruhemann continued in his post at Cambridge throughout the academic year 1914–1915 "with no thoughts of leaving," according to Martin D. Saltzmann, a professor at Providence College, Rhode Island, United States.[19] Unfortunately for him and his family, he began to be viewed with suspicion in some quarters even though he had lived in England for 30 years.

After the *Lusitania* sank life became even more difficult. The family doctor refused to retain them as patients, Martin Ruhemann was ostracised at school, and "hitherto congenial academic friends stopped visiting" the family, noted Sir Brian Pippard (1920–2008) in an article about Ruhemann published by the Royal Society in 1993—50 years after his death.[17] Pippard, who was professor of physics at Cambridge from 1971 to 1984, suggested that the *Cambridge Daily News* report in May 1915 probably led to Siegfried receiving threatening letters. He was sufficiently hurt to resign his lectureship at Cambridge. Even so, he remained a Fellow of the Royal Society and had just enough money to keep himself and his family in Cambridge until the end of the war.

The Ruhemanns returned to Berlin after the war. Siegfried initially served as an assistant in Fischer's laboratory and in 1921 was appointed head of an industrial research institute in Germany. His fellowship of the Royal Society lapsed in 1924, although the reason for this is not clear. His son Martin obtained a doctorate in physics at the University of Berlin in 1928. Siegfried retired in 1930. The family were Jewish, however. Martin emigrated to Britain to take up a research post at Imperial College, London, following the Nazi's rise to power in Germany in 1933.

The pogrom known as *Kristallnacht* or "Night of the Broken Glass" on 9–10 November 1938 when Nazi paramilitary forces and non-Jewish German civilians attacked Jews and looted their property finally persuaded Siegfried and Olga to leave Germany and join their son in London. They had retained their British nationality and returned to England in 1939 where they lived in Muswell Hill, a suburb of north London. Siegfried died from natural causes in August 1943 at the age of 84. "From 1919 to his death his British friends for the most part lost sight of him," remarks R. S. Morrell, his Chemical Society obituarist.[20] He died "so unobtrusively" that the Royal Society never gave him an obituary notice, "such as it has not denied other lapsed Fellows," Pippard observed.

POST-WAR COOPERATION AND FRIENDSHIP

Whereas Fischer had clamoured for the resumption of international scientific cooperation after the war, antagonism between scientists of the warring countries lingered for many years. "Under the conditions which obtained at the beginning of the twentieth century it is not surprising that some German chemists assumed the attitude that chemistry was a German science and that researches not published in German could be ignored," wrote American chemist William A. Noyes, Sr, (1857–1941) in a piece entitled "Criticism from Germany," published in 1922.[21] "To Americans such an attitude is very closely related to the attempt of Germany's political leaders to impose German imperialism on other countries by force. Any similar spirit on the part of Englishmen, Frenchmen or Americans, either in the political field or in science, is just as intolerable as it was in Germans."

Noyes was head of the chemistry department at the University of Illinois at Urbana-Champaign from 1907 to 1926 and ACS president in 1920. He was one of many chemists who tried to heal the wounds inflicted on international scientific cooperation by the war. In 1923, for example, he was a member of the United States delegation at the 4th general assembly of the International Union of Pure and Applied Chemistry (IUPAC) that was held in Cambridge, England. During this trip to Europe, he "spent much time ... trying to allay some of the hatreds between the French and Germans resulting from the war," observed another

Figure 18.7   William A. Noyes, Sr. (William Haynes Portrait Collection, Chemical Heritage Foundation Image Archives, Philadelphia, PA).

American chemist Roger Adams (1889–1971) in a biographical memoir of Noyes.[22]

It was through the efforts of chemists such as Noyes that Germany was eventually welcomed back into the international fold of chemistry. In 1930, IUPAC agreed to admit Germany as a member. Noyes served as IUPAC president from 1959 to 1963. The first German president of the union was inorganic chemist Wilhelm Klemm (1896–1985) who served from 1965 to 1967. Klemm obtained his PhD in chemistry under the supervision of Heinrich Biltz at the University of Breslau and then transferred to carry out research with his younger brother Wilhelm Biltz at Clausthal University of Technology. Wilhelm was a tank commander in World War I (see Chapter 15).

One of the most remarkable friendships to develop after the First World War was that between German chemist and Nobel laureate Fritz Haber and English physical chemist Harold Hartley. Haber had supervised the first German gas attack against the Allies in 1915 and became known as the "father of modern chemical warfare" (see Chapter 12).

Hartley was his counterpart in Britain. When war broke out, he was a lecturer in physical chemistry at Oxford University but "with the spirit of that time" immediately offered his services to the Royal Engineers.[23] He was turned down and instead sent to join the Leicestershire Regiment in Aldershot, a military town in Hampshire. With the advent of chemical warfare, he was summoned to the War Office in London and dispatched to France as a chemical adviser to the British 3rd Army. In 1917, he was promoted to lieutenant-colonel and appointed assistant director of gas services at the army's general headquarters in France. At the end of the war he was promoted to the rank of brigadier-general with responsibility for control of the Chemical Warfare Department at the Ministry of Munitions.

In the years after the war, Hartley was awarded the Military Cross and continued to advise the government on chemical weapons. He led a mission to Germany to inspect German chemical warfare research facilities and factories that had produced explosives and poison gases. As part of the trip, he paid a visit to Haber's Kaiser Wilhelm Institute for Physical Chemistry and Electrochemistry at Dahlem, Berlin, which had been the centre of chemical warfare research in the country. Haber and Hartley rapidly developed an amiable relationship, exchanging jokes and anecdotes about the use of gas in the war.[24] Hartley was knighted in 1928 and continued advising the British government on chemical warfare until 1950.

Haber similarly developed a friendly relationship with Cambridge University chemistry professor Sir William Pope who had headed the team that developed mustard gas in Britain (see Chapter 14). Pope arranged for Haber to stay in Cambridge following his forced emigration from Germany after the Nazis came to power in 1933. Other German chemists who were Jews followed Haber into exile in the late 1930s. They included Haber's friend Richard Willstätter, who also won the Nobel Prize in Chemistry, and Siegfried Ruhemann.

A GREAT CHEMIST WHO SURVIVED TWO WORLD WARS

It was not only in Germany that Jewish chemists who had been engaged in the First World War felt the full force of

Nazi power. Analytical chemist Fritz Feigl (1891–1971), who was born in Vienna, also suffered. Feigl studied chemical engineering in the city but when the First World War started he enlisted in the Austro-Hungarian Army. He served as an officer, was wounded in battle on the Russian front, and later received the Military Service Cross and other awards for his service.

When the war ended, Feigl resumed studies at the University of Vienna and obtained a doctorate in chemistry there in 1920. He became an assistant professor at the university the same year and was promoted to professor of inorganic analytical chemistry in 1935. Feigl is known mainly for his development of "spot-testing," a simple analytical technique that employs one or just a few drops of a chemical solution.

After Nazi Germany annexed Austria in March 1938 in what is known as the *Anschluss*, Fritz and his family emigrated to Switzerland and then Belgium. The Second World War erupted in September 1939 with the German invasion of Poland and in May the following year Germany attacked neutral Belgium. The Germans captured Feigl and transferred him to a concentration camp in the south of France which was then administered by the Vichy Regime.[25] His wife Regine and their son Hans, who were both away at the time of Fritz's capture, moved to France to be near the camp. With their help and that of the Brazilian Ambassador in Vichy France, they managed to obtain visas for all three of them to enter Brazil. The family arrived and settled in Rio de Janeiro as refugees in November 1940. The Brazilian authorities subsequently rescued Regine's two brothers who were also in a French concentration camp.

Fritz soon obtained a post as a researcher in *Laboratorio da Producao Mineral* (LPM), a mineral production laboratory in Rio de Janeiro, and around the same time began to teach a course on spot tests. He published a laboratory manual on the topic in 1943 and was granted Brazilian citizenship in 1944. Later in life, Fritz received many medals, prizes, and honours. He was awarded, for example, the Lmonosov Medal of the Academy of Sciences of Moscow and the Great Prize of Science and Culture of Austria. He was made an emeritus professor at the University of Tokyo and governor of the Hebrew University of Jerusalem, and also became a member of the board of governors of the Weizmann Institute in Rehovot, Israel.

The "Feigl Symposium" held by the Brazilian Academy of Sciences in Rio de Janeiro in November 1962 provided a good example of the high level of international cooperation and collaboration in science in general and chemistry in particular that had developed following the two world wars in the first half of the century. The symposium attracted eminent chemists from countries around the world including Austria, Germany, Israel, the United Kingdom and the United States.

One of the chemists was Ronald Belcher (1909–1982), professor of analytical chemistry at the University of Birmingham, England. "In terms of solid contribution and influence, Feigl ranks as one of the greatest analytical chemists of all time," he remarked in his obituary of Feigl.[26] "His knowledge of chemical reactions was unparalleled and his ability to exploit them for analytical purposes unsurpassed...His life's work—his legacy to chemistry—will be a lasting inspiration to all chemists." It was a fitting tribute to a chemist who was born in Austria, fought against the Russians and was wounded in the First World War, was forced into exile in Belgium when Nazi Germany annexed Austria, was interned in a concentration camp in France in the Second World War, and then became a refugee in Brazil.

REFERENCES

1. H. S. Raper, Arthur Smithells: 1860–1939, *Obit. Not. Fell. R. Soc.*, 1940, **3**, 97.
2. J. C. Bailar, Jr., Moses Gomberg: 1866–1947, *Biog. Mem. U.S. Nat. Acad. Sci.*, 1970, **41**, 141.
3. American Chemical Society, *A National Historic Chemical Landmark: The Discovery of Organic Free Radicals by Moses Gomberg*, American Chemical Society, Washington, DC, 2000.
4. W. Wetzel, Origins of and education and career opportunities for the profession of 'chemist' in the second half of the nineteenth century in Germany, in *The Making of the Chemist: The Social History of Chemistry in Europe 1789–1914)*, ed. D. Knight and H. Kragh, Cambridge University Press, Cambridge, 1998, p. 83.
5. W. Wetzel, Origins of and education and career opportunities for the profession of 'chemist' in the second half of the

nineteenth century in Germany, in *The Making of the Chemist: The Social History of Chemistry in Europe 1789–1914,)*, ed. D. Knight and H. Kragh, Cambridge University Press, Cambridge, 1998, p. 84.

6. T. Travis and T. Benfey, August Wilhelm Hofmann – a centennial tribute, *Education in Chemistry*, May 1992, p. 69.

7. G. Hartcup, *The War of Invention: Scientific Developments, 1914–18*, Brassey's Defence Publishers, London, 1988, p. 8.

8. J. W. Cobbs, Arthur Smithells: 1860–1939, *J. Chem. Soc.*, 1939, 1234.

9. Anon., Henry Stanley Raper (1882–1951), *J. R. Inst. Chem.*, 1952, **76**, 277.

10. W. P. Wynne, Obituary notices: Arthur William Crossley (1869–1927), *J. Chem. Soc.*, 1927, 3165.

11. War Office, Ministry of Munitions, Trench Warfare and Chemical Warfare Departments, and War Office, Chemical Warfare Research Department and Chemical Defence Experimental Stations (later Establishments), Porton: Reports and Papers, *WO 142*, 1905–1967.

12. A. W. Titherley, Obituary notices: James McConnan (1881–1916), *J. Chem. Soc., Trans.*, 1917, **111**, 316.

13. G. B. Kauffman, Emil Fischer: His role in Wilhelmian German industry, scientific institutions, and government, *J. Chem. Educ.*, 1984, **61**, 504.

14. G. B. Kauffman and P. M. Priebe, The Emil Fischer-William Ramsay friendship: The tragedy of scientists in war, *J. Chem. Educ.*, 1990, **67**, 93.

15. W. Ramsay, The 'Zeppelin' scare, *The Times* [London, England] 24 Oct. 1914, p. 7.

16. S. Highly-place spies, *The Times* [London, England] 22 Aug. 1914, p. 7.

17. B. Pippard, Siegfried Ruhemann (1859–1943), F.R.S. 1914–1923, *Notes Rec. R. Soc. Lond.*, 1993, **47**, 271.

18. R. West, Siegfried Ruhemann and the discovery of ninhydrin, *J. Chem. Educ.*, 1965, **42**, 387.

19. M. D. Saltzmann, Is science a brotherhood? The case of Siegfried Ruhemann, *Bull. Hist. Chem.*, 2000, **25**, 116.

20. R. S. Morrell, Siegfried Ruhemann, 1859–1943, *J. Chem. Soc.*, 1944, 40.

21. W. A. Noyes, Criticism from Germany, *Ind. Eng. Chem.*, 1922, **14**, 99.
22. R. Adams, *William Albert Noyes: 1857–1941: A biographical memoir*, National Academy of Sciences, Washington, DC, 1952.
23. A. G. Ogston, Harold Brewer Hartley. 1878–1972, *Biogr. Mems. Fell. R. Soc.*, 1973, **19**, 349.
24. D. Charles, *Between Genius and Genocide: the Tragedy of Fritz Haber, Father of Chemical Warfare*, Pimlico, London, 2006, p. 181.
25. A. Espinol, M. Abrantes da Silva Pinto and C. Costa Neto, Fritz Feigl (1891–1971). The centennial of a researcher, *Bull. Hist. Chem.*, 1995, **17/18**, 31.
26. R. Belcher, Obituary: Fritz Feigl, *Proc. Soc. Anal. Chem.*, 1971, **8**, 172.

CHAPTER 19

One Building, Two Memorials

HOW WILL I BE ACCOUNTED FOR?

I cannot recall on how many occasions I've visited the London offices of the Royal Society of Chemistry (RSC) over the past few decades. It could possibly be dozens if not scores of times. I know the RSC Library and Information Centre well and have regularly attended meetings and functions in the same building. Yet I'm ashamed to say that it was not until I started to carry out research for this book that I became aware of two memorials in the building commemorating chemists who died on active duty during the First World War.

The building is located in the east wing of Burlington House on Piccadilly. Its neighbour, the Royal Academy of Arts, occupies the north wing at the far end of the courtyard. As you step from the courtyard into the RSC building through heavy wooden doors, it is easy to miss the bronze-on-oak Royal Institute of Chemistry (RIC) war memorial on the right before you arrive at the reception desk.

The Institute of Chemistry of Great Britain, which subsequently became the Royal Institute of Chemistry, was founded in 1877 to serve the chemistry profession. In 1980, the institute merged with three other bodies: the Chemical Society, a learned society created in 1841; the Society for Analytical Chemistry

The Chemists' War, 1914–1918
By Michael Freemantle
© Freemantle, 2015
Published by the Royal Society of Chemistry, www.rsc.org

Figure 19.1   Royal Society of Chemistry, Burlington House, London (author's own, 2013).

founded in 1874; and the Faraday Society named after the famous English chemist and physicist Michael Faraday (1791–1867) and founded in 1903 to advance the study of physical chemistry. The combination of the four separate bodies became the Royal Society of Chemistry with the twin aims of serving as a professional body and a learned society.

The RIC memorial is dedicated to the "Fellows, Associates and Students of the Institute of Chemistry who died in the service of their country 1914–1918." It has three columns listing a total of 55 names. As part of my research, I decided to delve into the backgrounds of a random sample of these chemists.

The name at the top of the left hand column is James Watson Agnew. I discovered that he served as a lieutenant in the Highland Light Infantry and was killed in action in France on 21 May 1915. His name also appears on the Le Touret Memorial in the Pas-de-Calais region of France. The memorial lists over 13 000 British and Commonwealth soldiers who were killed in the region and who have no known graves.

On the same day that I was looking into Agnew's record, I read the poem "Soliloquy" by Irish war poet Francis Ledwidge (1887–1917) who was killed by a shell on the first day of the Battle of Passchendaele fought in Belgium from 31 July to 10 November 1917.[1] Whether Agnew drank wine or not while in France, I do not know, but the following lines from the poem seemed apt:

And now I'm drinking wine in France,
The helpless child of circumstance.
Tomorrow will be loud with war,
How will I be accounted for?

So how did posterity account for Agnew? Most of the names on the RIC memorial can be found online on the RSC Bibliographic Database.[2] The entry for Agnew gives his year of death (1915) but, unlike many other entries on the database, not his year of birth. However, the entry does provide a link to an issue of the *Proceedings of the Institute of Chemistry of Great Britain and Ireland* that was published in 1915.[3] A short obituary about Agnew in that issue notes that before the war he lectured in chemistry in Glasgow, first at the Royal Technical College and then at St Mungo's College. He was made "an Examiner" in chemistry to the Glasgow Royal Faculty of Surgeons and Physicians in 1909, elected a Fellow of the institute in 1910 and co-authored a book on practical chemistry published in 1911.[4] My investigations revealed little else about Agnew.

The same issue of the institute's proceedings also has brief obituaries of eleven other chemists including two who, like Agnew, died while on active service in 1915 and whose names appear on the RIC memorial. Joseph Walter Harris died on 3 June at the age of 28. After graduating at University College, Nottingham in 1914, he qualified as an associate of the institute and was appointed "Chemist" at Shirebrook Colliery, near

Figure 19.2   Royal Institute of Chemistry memorial (author's own, 2013).

Mansfield in England, a position he held until the outbreak of the war. He was serving as a lieutenant in the Lincolnshire Regiment at the time of his death. His name is among the 54 896 names of British and Commonwealth soldiers with unknown graves that appear on the Menin Gate Memorial to the Missing at Ypres, Belgium.

The other chemist is Thomas Wright (1887–1915). Like Harris, he was 28 when he was killed in action. Wright graduated with a first class honours degree in chemistry from King's College, London, in 1912 and in the same year qualified as an associate of the institute. According to his obituary in the proceedings, Wright was preparing "to go to Germany for further study" when the war started. He died in France on 2 May 1915. Thomas Wright is the last of the 55 names listed on the RIC memorial.

ASHAMED TO WALK ABOUT THE CITY

As you pass the reception desk in the RSC's Burlington House and climb the stairs on the way to the library, you see an

Figure 19.3   Chemical Society memorial (author's own, 2013).

imposing bronze and marble memorial that commemorates 30 members of the Chemical Society who died during the Great War. Their names are inscribed in the marble below the Latin words *"PRO PATRIA* (Figure 19.3)."

William Saunders (1889–1916) is one of four chemists whose names appear on both the Chemical Society and RIC memorials. Obituaries of all four were published in the *Journal of the Chemical Society, Transactions* as well as in the RIC proceedings during the war.

Saunders, a tall good-looking young man, attended school in the city of Liverpool, England, and studied in Germany and Switzerland. He was the only son of W. H. Saunders, chairman of Ayrton Saunders, a Liverpool-based company that manufactured pharmaceuticals and related products. After working in the company for two years, William took courses at the Pharmaceutical Society of Great Britain's college in London (now UCL School of Pharmacy and part of University College London) and Liverpool University. He subsequently qualified as an associate of the RIC and was also elected a Fellow of the Chemical Society. By the time the war started, the young Saunders had become a director of Ayrton Saunders.

"Immediately on the outbreak of war he placed himself at the service of the Government, thinking possibly his chemical knowledge and experience might be utilised," noted his father in a two-page obituary of his son published in the society's journal.[5] "After waiting, however, for about three months without receiving any replies except acknowledgements of the applications sent by him, and feeling, to use his own words, 'ashamed to walk about the city,' he joined the 5th King's Liverpool Rifles as second lieutenant, afterwards becoming lieutenant and adjutant and, in July, 1916, captain and adjutant of the battalion."

William Saunders was killed in action in France on 6 September 1916 during the Battle of the Somme and is buried near Albert at the Dartmoor Cemetery, Becordel-Becourt.

The name of Canadian Norman Campbell (1886–1917) also appears on both memorials in Burlington House. Campbell was born in Chicago but received his early education in Montreal and at Dulwich College, England. He then proceeded to Balliol College, Oxford University, graduating with a first class honours degree in chemistry. In 1907, he travelled to Ceylon (now Sri Lanka) to take up a missionary post as Professor of Science at Trinity College, Kandy. On the way there, he carried out an investigation of the salinity of the Indian Ocean.

Campbell returned to England in December 1914 and enlisted in the British Army. In April 1915, following the German chlorine gas attack at Ypres in Belgium, he volunteered to join one of the four Special Companies, later to become known as the Special Brigade, that the Royal Engineers established in May 1915 to take responsibility for waging chemical warfare. He was wounded during the first British gas attack at the Battle of Loos in September 1915 and killed near Arras by a German machine gun bullet on 3 May 1917. An officer with him laid his body in a shell hole and remained with it for three hours but was later forced to

FIRST-AID ON THE BATTLE-FIELD: A CAMERA IMPRESSION ON THE SOMME FRONT. [*British official photograph.*]
Wounded soldiers undergoing quick but expert treatment at an advanced dressing-station.   As soon as their immediate needs were satisfied the motor-ambulance conveyed them to the base hospital. On the horizon the smoke of a bursting shell can be distinctly seen.

Figure 19.4   First aid at an advanced dressing station on the Somme front where Saunders was killed (Wellcome Library, London).

abandon the body when the Germans attacked.[6] The attack subsequently made the spot inaccessible and so it was impossible to recover his body. Campbell is commemorated on the Arras Memorial in the Faubourg d'Amiens British Cemetery. The memorial bears the names of 34 785 British and Commonwealth soldiers who were killed in the area during the war and whose bodies were never found.

HOW HELPLESS YOU FEEL

Obituaries for most of the names on the Chemical Society memorial appear in volumes of the *Journal of the Chemical Society, Transactions* published between 1915 and 1919. The 1915 volume, for example, has obituaries for just two chemists who were killed in action during the war. Research chemist John Dunlop, born in Belfast in 1885, was given a commission in the Royal Dublin Fusiliers at the start of the war.[7] He was killed on

27 August 1914 during the "Great Retreat" of the British Expeditionary Force after their first major engagement in Belgium, the Battle of Mons on 23 August.

Cecil Crymble, like Dunlop, was also a research chemist and born in Belfast in 1885.[8] He was attached to the Royal Irish Fusiliers while serving in France and, when he could, loved to write highly descriptive letters home.[9]

In one of his last letters, dated 12 November 1914, he wrote (from an unspecified location): "The worst experience I have met with yet occurred yesterday when I was on my way here with a party from the trenches. We had to go through a small town which was being shelled; we could see them putting a couple every minute on either side of the road we had to go along, and most times a house would go up in the air. Finally one lit 50 yards in front of us. Problem–were they going to put the next one further along the road, or behind us? Fortunately it arrived behind us and short. We lost no time getting along the road. It is extraordinary how helpless you feel. You can hear the whiz of it as it comes, but there is no use moving, as you do not know which way to go."

He adds: "Life at Divisional Headquarters is very serene. Hearty meals are consumed, undeterred by news that the enemy are attacking strongly on the right or left." On 20 November 1914, just over a week after he wrote this letter, he was killed by a sniper's bullet in trenches near the French town of Armentières which lies on the border with Belgium. His name appears on the Ploegsteert Memorial to the Missing that bears the names of over 11 000 other British and Commonwealth soldiers who died on the Ypres Salient on the Western Front and who have no known graves. The memorial is situated on the road between Armentières and Ypres.

As you look into the life and death of chemists such as Dunlop and Crymble and as you dig into the backgrounds of other chemist-soldiers on First World War memorials, it is not difficult to envisage the personal pain, grief, and loss behind each name. All these young men sacrificed their lives in the war and at the same time their countries forfeited an immense amount of talent, experience and energy. Each loss was a tragedy.

At the annual meeting of American Chemical Society held in Philadelphia in September 1919, Newton D. Baker (1871–1937), US Secretary of War, alluded to such loss in a speech entitled "Chemistry in Warfare."[10] The death of so many strong and virile men in the flower of their youth resulted in not just "orphanage and widowhood," he said, but also in the withdrawal of "vast energy from the productive forces of civilization."

The obituary notices of the *Journal of the Chemical Society* published during and just after the war provide ample evidence of the loss of this productive energy in chemistry, and no more so than the loss to the teaching profession. The first eight obituaries in the 1918 volume of the journal are devoted to seven chemists killed in action in 1917 and one killed in 1915.[11] All but two of the eight chemists taught science at one level or another before the war and then joined the British Army, mostly at the outbreak of the war or soon after.

Arthur Edwin Tate (1880–1917), whose name appears on the Chemical Society memorial, is typical. After graduating with a bachelor's degree in chemistry in 1901, he taught science at various schools in the north of England. "In 1910 he was appointed senior science master of the City of Norwich School, a position which he filled with marked success," according to his Chemical Society obituarist.[12] "He joined the Royal Engineers in July, 1916, and went to France in the following September, where he was wounded on April 12th, 1917, and died from the effects of his wounds ten days later."

One can only imagine the suffering he endured during those ten days and the shock and sorrow of his wife and family after they received the news of his death. "Arthur E. Tate was a man of sterling qualities, a good and successful teacher," his obituary concludes. "He leaves a widow, to whom he was married shortly before he joined the Army."

SCIENCE TERRIFIES, BUT SCIENCE ALSO PROTECTS

At the top of the Chemical Society war memorial is a bronze plaque showing three soldiers on watch armed with rifles. One of them is about to sound the gas alarm, a bell. The surrounding inscription reads: "To save our armies from poison gas" and "He gave the last full measure of devotion." The plaque is dedicated to Edward Frank Harrison (1869–1918) who died during the war but, unlike most of the other chemists commemorated on the two memorials, was not killed in action (Figure 19.5).

Figure 19.5   Plaque dedicated to Edward Harrison (author's own, 2013).

Whereas many chemists on active duty in the British Army did not have the opportunity to apply their expertise directly to the war effort, Harrison's knowledge and experience of chemistry had a major impact. Together with a team of co-workers, he was responsible for the design and development of respirators that were used in their millions by British and Allied troops during the war for protection against gas attacks.

Harrison was born in London, trained as a pharmaceutical chemist, became a Fellow of the Chemical Society in 1894, and was elected a Fellow of the Institute of Chemistry in 1905.[13] He is one of the four chemists whose names appears on both memorials in Burlington House. When war broke out, he repeatedly tried to enlist but was refused each time on the grounds of his age. He was 45 at the time. Then, in May 1915, he managed to enlist as a private but was soon transferred to the Royal Army Medical College in London to work in the anti-gas service which had been set up after the first German chlorine attack in April 1915.

The earliest attempts to protect against chlorine following the attack were improvised and fairly primitive (see Chapter 13). As chemical warfare accelerated with further chlorine attacks by the Germans and then by the British as well, and with the introduction of other poison gases such as phosgene, anti-gas measures rapidly developed on both sides. The Germans introduced a snout-type canister gas mask at the end of 1915. The canister contained neutralising chemicals to protect against chlorine and also materials such as activated charcoal that adsorbed phosgene and other toxic gases.

The respirator designed by Harrison and his team, known as the British "Large Box Respirator" or "Harrison's Tower," was the first British respirator to employ activated charcoal. It consisted of a muslin facemask impregnated with a solution containing hexamine, an organic compound that counteracted phosgene. A corrugated rubber breathing tube connected the facemask to a separate filter canister that was carried in a haversack. The canister contained activated charcoal, soda lime—a strongly alkaline mixture of calcium hydroxide and a small amount of sodium hydroxide—that neutralised acidic gases such as chlorine, and potassium permanganate, a powerful oxidising agent that destroyed both lethal gases and tear gases.

Harrison's respirator was first used by the British Army in February 1916 and by June that year around 200 000 had been issued to troops. The respirator proved heavy and unwieldy, however. By August 1916, a more compact version, the British Small Box Respirator, had become standard issue in the army. Harrison is said "to have been responsible for the organisation for supplying the British and certain of the Allied Armies with nearly fifty million respirators to protect against gas attacks," notes his obituary in the institute's proceedings.[14]

His work during the war took him to France many times to meet with chemists on the front. In October 1918, the French made him officer in the Legion d'Honneur, their highest military honour.

While working in London as Controller of the Chemical Warfare Department responsible for anti-gas measures, Harrison fell victim to influenza and died from pneumonia on 4 November 1918, one week before the Armistice was signed. He was 49 years old and about to be promoted to the rank of

Figure 19.6   British box respirator (author's own, 2012, Imperial War Museums, London).

brigadier-general, having previously held the rank of lieutenant-colonel in the Royal Engineers.

As part of his work, "he continuously exposed himself to toxic chemicals but ironically became one of the victims of the influenza epidemic in November 1918," writes British author Guy Hartcup in a book about scientific developments during the war.[15] The epidemic of "Spanish flu," as it was called, probably originated in the United States and then quickly spread around the world becoming a pandemic. The first outbreaks of the disease in the British Army in France occurred in April and May 1918.[16] According to some estimates, by the end of the pandemic between 50–100 million people around the world had died from the disease.[17]

Harrison was not the only eminent member of the British Army to die prematurely from Spanish flu. Colonel Sir Mark Sykes

(1879–1919), for example, died from the disease in Paris on 16 February 1919 at the age of 39. Sykes worked with French diplomat François George-Picot (1870–1951) to draw up plans for the British and French governments to dismember the Ottoman Empire and divide up its Arab territories in the Middle East between Britain and France. The secret agreement, concluded in May 1916, became known as the Sykes–Picot agreement.

Harrison's obituary, a four-page notice written by "A. S." and "H. S. R.," followed in the 1919 volume of the Chemical Society's journal.[18] "His bodily strength, sapped by unceasing labour, was unequal to the strain, pneumonia supervened, and he passed away, as certainly as anyone on the battlefield, a sacrifice of the war," it explains, and then adds, sadly: "His elder son had fallen in France in 1916." Aged just 19, he was killed in action on 30 July during the Battle of the Somme.

Edward Harrison was buried with full military honours at Brompton Cemetery, London. After his death, Winston Churchill, who was Minister of Munitions at the time, wrote to Harrison's widow, Edith, saying "it is due in large measure to him that our troops have been given effectual protection from the German poisonous gases."[19]

The Chemical Society memorial to Harrison and other Fellows of the Chemical Society was unveiled on 16 November 1922 by David Lindsay (1871–1940) who was the 27th Earl of Crawford and 10th Earl of Balcarres.[20] Lindsay had a distinguished career. After graduating in history from Oxford University in 1894, he pursued a career in politics and at the same time took a keen interest in scientific research.[21] Early on in the First World War he enlisted as a private in the Royal Army Medical Corps (RAMC) and served as Corporal Crawford for over a year. He was made First Commissioner of Works in the British government in April 1921 and Minister of Transport in April the following year. He retained both posts until October 1922. In 1924, the Royal Society, Britain's most prestigious scientific society, made him a Fellow of the society.

During the unveiling ceremony at Burlington House, Sir James Walker (1863–1935), a Scottish chemist who was president of the Chemical Society from 1921 to 1923, called upon Lindsay to speak.

"The name of Edward Harrison stands out prominently amongst those of a gallant band called upon to protect, indeed to save our forces from the most terrible assaults ever directed by science against an army in the field," Lindsay said. "During the spring of 1915 our troops found themselves faced by the horrors of an unknown and inconceivable style of warfare. Science terrifies, but science also protects."

At the conclusion of the unveiling ceremony Walker thanked Lindsay and noted that, as First Commissioner of Works, he had been responsible "for the fabric in which the Memorial stands." He pointed out that Lindsay had enlisted in the RAMC "and saw with his own eyes the ravages of poison gas and the success of the protective measures employed against it."

THE SILVERTOWN EXPLOSION

There are two columns of 15 names inscribed under the words *"PRO PATRIA"* on the Chemical Society war memorial. The right-hand column is headed by Edward F. Harrison and the left column by Andrea Angel (1877–1917). Harrison and Angel are similar in several ways. They were both chemists who contributed their talents in chemistry to the war effort, neither was killed in action, and both saved lives, albeit in different ways.

However, there was a major difference in the work of the two men. Whereas Harrison focused on the "science also protects" aspect of chemistry, Angel devoted himself to the "science terrifies" side of the subject. He applied his skills to the manufacture of military explosives during the war.

Born in Bradford, Angel graduated with a first class honours degree in chemistry from Oxford University in 1899. When the war started, he was a tutor and lecturer in chemistry responsible for running the laboratory at Christ Church, one of the university's colleges.[22] After volunteering to contribute his expertise in chemistry for war work, he was appointed chief chemist at a munitions factory owned by the chemical manufacturing firm Brunner Mond. The factory was located in Silvertown, a densely populated industrial area east of London on the north bank of the River Thames.

The company had opened the chemical works in 1894 to produce sodium carbonate (washing soda) and sodium hydroxide (caustic soda). Despite opposition from Brunner Mond because of the population density around the works, the British War Office decided to use the factory for the manufacture of TNT. Production started in September 1915.

"At the beginning of the war, the manufacture of TNT was poorly understood," observes archaeological investigator Wayne D. Cocroft in an account of explosives manufacture in Britain during War I.[23] The flammability of the material was thought to be the principle danger.

On 19 January 1917, a fire broke out in one of the upper rooms of the Silvertown factory. Angel attempted to put out the fire. "Realising that a fire in such a place could scarcely be stopped, all Angel's care was given to getting the workers, largely women and girls, into a place of safety," according to his obituary in the Chemical Society's journal.[24]

The fire ignited about 50 tons of TNT, much of it stored in nearby railway wagons waiting to be hauled out of the factory. The explosion demolished a substantial part of the factory as well as houses for miles around. It killed 73 people, including Angel, and injured hundreds more. The blast was heard on the south coast of England over 100 miles away and the fires it started were seen some 30 miles way.

Two months later, on 24 March, King George V posthumously awarded the Albert Medal for Lifesaving, first class, to Angel for his bravery. He presented the medal to his widow, Mary. After the Second World War, the Albert Medal award was replaced by the George Cross.

The Silvertown TNT explosion was the worst explosion in and around London during the war in terms of the number of casualties, but there were even worse explosions in other parts of England. On 2 April 1916, for example, around 15 tons of TNT and 150 tons of ammonium nitrate exploded at the Explosives Loading Company factory near Faversham in Kent. Known as "the Great Explosion," the blast killed over 100 people, destroyed five buildings and created a massive crater. It was heard across the English Channel in France. On 1 July 1918, an explosion of eight tons of TNT at the government-owned National Shell Filling Factory at Chilwell, a village in Nottinghamshire, demolished much of the factory, killed 134 people, and injured another 250.

CANNOT BE REPLACED

"Under the exigency of the moment, both England and France allowed men to go to the front at the outbreak of the war who should have been retained at home at all costs; such men as Mosely (*sic*), for example, cannot be replaced, and both France and England are feeling at the present time the loss of the scientific experts who went to the front." So wrote Marston Taylor Bogert, chairman of the United States National Research Council Chemistry Committee in a report published in November 1917.[25]

Bogert continues: "As soon as the situation permitted, both countries sent thousands of scientists and skilled artisans back from the front to the aid of the industries at home, realizing very quickly that modern warfare depends absolutely upon industrial organization and efficiency. It has been stated upon creditable authority that England alone has recalled from the fighting line over 128 000 of such men."

Henry Moseley (1887–1915), who made an immense contribution to chemistry before the war, was not recalled. Like most of the chemists listed on the Burlington House memorials, he was killed in action. On 10 August 1915, he was shot in the head by a sniper's bullet at Suvla Bay, Turkey, during the Gallipoli campaign against the Ottoman army. He is not listed on either memorial in Burlington House, and for good reason. He was a not a chemist, but a physicist.

Born in Dorset, England, Moseley graduated from Oxford University in 1910 and subsequently carried out research in physics at the University of Manchester, working with physicist Ernest Rutherford who had won the Nobel Prize in Chemistry in 1908. Three years after his prize, Rutherford discovered the atomic nucleus, the centre of an atom that contains particles with a positive electric charge, known as protons, and electrically neutral particles, known as neutrons.

In November 1913, Moseley moved from Manchester to Oxford to pursue his research in the university's laboratories and then, in the summer of 1914, he decided to apply for the post of

Professor of Physics at the University of Birmingham,[26] but then the war intervened. He volunteered his services, was commissioned in the Royal Engineers as a brigade signal officer with the rank of second lieutenant, and sent to the Dardanelles on 13 June 1915.

By the time the war started, he had published just eight papers, the last two of which, published in 1913 and 1914, were to have a far-reaching impact on science in general and chemistry in particular. Anyone who has ever studied the Periodic Table of the Chemical Elements will know that each element has a number. Hydrogen, for example, is numbered 1, carbon is 6, oxygen 8, and iron 26. These numbers are fundamental properties of the atoms of which the elements are composed. They distinguish one element from another and correspond to the number of protons in each atom of the element. Each atom of hydrogen has one proton in its nucleus; the nucleus of a carbon atom has six protons, and the nuclei in oxygen and iron atoms have 8 and 26 protons, respectively.

Moseley discovered these numbers, or integers as he called them, by using photographic techniques to analyse the X-rays emitted by solid elements when they are bombarded by fast streams of cathode rays. "We can therefore conclude from the evidence of the X-ray spectra alone, without using any theory of atomic structure, that these integers are really characteristic of the elements," he wrote in his 1913 paper reporting the discovery.[27] He noted that it was improbable that two different stable elements should have the same integer and concluded that "the integer which controls the X-ray spectrum is the same as the number of electrical units in the nucleus," the electrical units being protons. Moseley called these integers "atomic numbers."

By means of these studies, he was able to predict the existence of three elements that were missing from the periodic table. "Known elements correspond with all the numbers between 13 and 79 except three," he concluded in his 1914 paper. "There are here three possible elements still undiscovered."[28] He was referring to elements with atomic numbers 43, 61, and 75, later to be identified as technetium, promethium, and rhenium, respectively.

"Had Moseley lived, he would probably soon have received the Nobel prize for physics or chemistry," suggests American historian of science John L. Heilbron.[29]

THOSE WHO SURVIVED

It is probably difficult to determine how many chemists fought in the war and survived. Many who did survive never rose to fame and were unknown by the scientific community at large or, if they were known, they were soon forgotten. But they were not forgotten by their families.

Horace Batkin (1895–1931) is just one example of such a chemist and his story is remarkable particularly as he served gallantly in the war from beginning to end. His niece, Angela Powell, provided me with the details. Batkin was born and educated in Burton-upon-Trent, a town in Staffordshire, England. Soon after leaving school he trained as an apprentice in a chemist's shop in the town. He enlisted in the army on 5 August 1914, the day after Britain declared war on Germany, and was mobilised the following month. He saw active service as a sapper in a signal company of the Royal Engineers in France and then at the end of the war he served for a few months in Egypt.

While in France, Batkin took part in the Battle of Hohenzollern Redoubt, one of the battles of the Battle of Loos fought from 25 September to 15 October 1915, and the following year was in action again at Neuville-St-Vaast, a village near Arras in northern France. He was one of the earliest recipients of the British Army's Military Medal. The medal was established in March 1916 to award troops below commissioned rank for their bravery in battle.

The citation for his medal reads: "During the attack on the HOHENZOLLERN REDOUBT 13th October, 1915, the lines under the charge of this man were heavily shelled, with the result that he had practically to patrol them continuously. At NEUVILLE-ST-VAAST in March and April, 1916, in the same capacity during frequent bombardment of communication trenches he did excellent work and showed great resolution."

Batkin was in action again at the Battle of St Quentin from 21–23 March 1918. The battle was fought at the beginning of Operation Michael, a series of German attacks that launched their Spring Offensive against the Allies in northern France. In the first five hours of the attack, the Germans bombarded Allied positions with over one million high explosive shells and shells filled with chemical warfare agents such as mustard gas. Batkin was badly gassed during the attack and suffered from the after-effects for some two years.

Following demobilization early in 1919 he took up a job with a wholesale chemists firm in Birmingham where he was put in charge of the "chemical room." He died at the age of 36 from illness which, it was thought, was caused by gassing during the war.

REFERENCES

1. *The War Poets*, (poems of the First and Second World Wars selected by Michael Wylie), Pitkin Publishing, Andover, 1992, p. 59.
2. The RSC Biographical Database, 27 June 2014, http://www.rsc.org/Library/Services/GenealogyEnquiriesSearch.asp.
3. Anon., Obituary. James Watson Agnew, *Proc. Inst. Chem. GB Irel.*, 1915, **39**, C22.
4. G. B. Neave and J. W. Agnew, *An Introduction to Practical Chemistry*, Blackie & Son, London, 1911.
5. W. H. Saunders, William Gilbert Saunders, *J. Chem. Soc., Trans.*, 1917, **111**, 320.
6. Memorial Plaque: Norman Phillips Campbell, 22nd November 2013, http://www.northeastmedals.co.uk/foreign_war_medals_miscellaneous/memorial_plaque_norman_phillips_campbell.htm.
7. T. B. W. John, Gunning Moore Dunlop, *J. Chem. Soc., Trans.*, 1915, **107**, 582.
8. T. B. W. John, Gunning Moore Dunlop, *J. Chem. Soc., Trans.*, 1915, **107**, 581.
9. Extracts from Letters of the Late Lieut. C. R. Crymble, D.Sc., Royal Irish Fusiliers, 22nd November 2013, http://www.instgreatwar.com/page31.htm.
10. N. D. Baker, Chemistry in warfare, *Ind. Eng. Chem.*, 1919, **11**, 921.
11. Anon., Obituary notices, *J. Chem. Soc., Trans.*, 1918, **113**, 300.
12. Anon., Obituary notices: Arthur Edwin Tate, *J. Chem. Soc., Trans.*, 1918, **113**, 306.

13. Anon., Obituary: Lieut.-Colonel Edward Frank Harrison, *Proc. Inst. Chem. GB Irel.*, 1918, **42**, D25.
14. Anon., Obituary: Lieut.-Colonel Edward Frank Harrison, *Proc. Inst. Chem. GB Irel.*, 1918, **42**, D26.
15. G. Hartcup, *The War of Invention: Scientific Developments, 1914–18*, Brassey's Defence Publishers, London, 1988, p. 103.
16. Anon., The influenza epidemic in the British armies in France, 1918, *Br. Med. J.*, 1918, **2**, 505.
17. J. M. Barry, *The Great Influenza: The Story of the Deadliest Pandemic in History*, Penguin, London, 2004, p. 4.
18. A. S. and H. S. R., Edward Frank Harrison, *J. Chem. Soc., Trans.*, 1919, **115**, 562.
19. Gas mask inventor Harrison honoured in death by Churchill, RSC press release, 2008, http://www.rsc.org/AboutUs/News/PressReleases/2008/HarrisonLetter.asp.
20. Anon., War Memorial, *J. Chem. Soc., Trans.*, 1922, **121**, 2894.
21. W. H. Bragg, David Lindsay Earl of Crawford and Balcarres 1871–1940, *Obit. Not. Fell. R. Soc.*, 1941, **3**, 404.
22. Andrea Angel BSc MA: Edward Medal (1st Class), 27th November 2013, http://www.chch.ox.ac.uk/cathedral/memorials/WWI/Andrea_Angel.
23. W. D. Cocroft, First World War Explosives Manufacture: The British Experience, in *Frontline and Factory: Comparative Perspectives on the Chemical Industry at War, 1914–1924*, ed. R. MacLeod and J. A. Johnson, Springer, Dordrecht, 2006, p. 35.
24. H. B. B. Andrea Angel, *J. Chem. Soc., Trans.*, 1917, **111**, 321.
25. M. T. Bogert, National Research Council Chemistry Committee Second Report, *Ind. Eng. Chem.*, 1917, **9**, 1009.
26. J. L. Heilbron, The work of H. G. J. Moseley, *Isis*, 1966, **57**, 336.
27. H. G. J. Moseley, The high frequency spectra of the elements, *Phil. Mag.*, 1913, **26**, 1024.
28. H. G. J. Moseley, The high frequency spectra of the elements, Part II, *Phil. Mag.*, 1914, **27**, 703.
29. J. L. Heilbron and H. Moseley, and the Nobel prize, *Nature*, 1987, **330**, 694.

Fifty Chemicals of the Great War

The death and destruction of the First World War and the care of the sick and wounded in the four-year conflict would not have been possible without the production of a vast variety of chemicals. The following is a selection of fifty of the more common chemicals and chemical products that were used in the war. Many are listed under names used at the time and in the general non-technical and fiction literature of the First World War, rather than under their chemical names. For example, the organic compound with the formula $(ClCH_2CH_2)_2S$ is listed under mustard gas rather than under dichloroethyl sulfide or one of its other chemical names.

Acetone, $(CH_3)_2CO$. A solvent with numerous wartime uses including the manufacture of cordite, cellulose acetate, and tetryl.

Activated charcoal. A highly porous form of carbon used in gas masks to adsorb toxic gases.

Aluminium, Al. A light metallic chemical element widely used in the war, notably for the manufacture of alloys needed to make the airframes of Zeppelins and other aircraft (Figure 20.1). It is extracted from an ore known as bauxite which consists principally of aluminium hydroxide, $Al(OH)_3$.

Amatol. A high explosive mixture of 60% ammonium nitrate and 40% TNT used by the British to fill shells.

The Chemists' War, 1914–1918
By Michael Freemantle
© Freemantle, 2015
Published by the Royal Society of Chemistry, www.rsc.org

Figure 20.1 Zeppelin emerging from its hangar. An aluminium alloy was used to make its frame. (Wellcome Library, London).

Ammonal. The name given to a variety of high explosives containing ammonium nitrate and aluminium. It was used as a filling for hand grenades, shells, and trench mortar bombs. The British used ammonal as a blasting explosive in tunnelling operations to blow up mines under German lines on the first day of the Battle of the Somme, 1 July 1916.

Ammonia, NH_3. A pungent gas that dissolves in water to form the alkali ammonium hydroxide, NH_4OH. It was manufactured in Germany in the war from nitrogen, N_2, (extracted from air) and hydrogen, H_2, using an iron-based catalyst:

$$N_2 + 3H_2 \rightarrow 2NH_3$$

The process is known as the Haber–Bosch process after the two German chemists who developed it. Ammonia was used to make fertilisers such as ammonium sulfate, $(NH_4)_2SO_4$, and the nitric acid required to manufacture explosives such as TNT.

Ammonium nitrate, NH_4NO_3, was prepared by the reaction of ammonia and nitric acid.

It was used as a fertiliser and as a component in explosive mixtures such amatol and ammonal.

Aspirin, also known as acetylsalicylic acid, is an analgesic and antipyretic drug.

Aspirin was given to sick and wounded soldiers in the war to relieve pain and reduce inflammation and fever.

Ballistite. A British smokeless propellant consisting of a mixture of nitroglycerine and collodion in equal proportions.

Boric acid, H_3BO_3, also known as boracic acid, was used as an astringent and antiseptic in preparations such as Dakin's solution and eusol.

Bronze. A copper alloy that typically contains 80% copper and 20% tin, but may also contain one or more other elements such as aluminium, iron, manganese, and phosphorus. The *Lusitania*, sunk off the south coast of Ireland by a German U-boat on 7 May 1915, was propelled by four 15 ton manganese bronze propellers. Brass, another copper alloy, typically contains 80% copper and 20% zinc.

Calcium cyanamide, $CaCN_2$, also known as nitrolime, is a white powder that was used as a nitrogenous fertiliser and also, by its reaction with water, to make ammonia for the war effort:

$$CaCN_2 + 3H_2O \rightarrow CaCO_3 + 2NH_3$$

$CaCN_2$ was manufactured by heating calcium carbide, CaC_2, in nitrogen, N_2. Calcium carbide was produced by heating a mixture of lime (calcium oxide, CaO) and coke (carbon, C). Lime was made by heating limestone (calcium carbonate, $CaCO_3$). The reaction sequence starting with limestone was:

$$CaCO_3 \rightarrow CaO + CO_2$$

$$CaO + 3C \rightarrow CaC_2 + CO$$

$$CaC_2 + N_2 \rightarrow CaCN_2 + C$$

Carbolic acid, also known as phenol, was used in the war as a disinfectant and antiseptic.

It was obtained either directly by distilling coal tar, or indirectly from another coal tar chemical, namely benzene, C_6H_6. In the latter process, the benzene was first treated with sulfuric acid and the product of the reaction then fused with caustic soda (sodium hydroxide, NaOH).

Cellulose acetates are synthetic polymers with complicated chemical structures. They are prepared from cellulose—the main constituent of plant fibre. The polymers were used in the war to prepare non-flammable photographic films and fire-resistant lacquers to waterproof and stiffen aircraft fuselage (Figure 20.2).

Figure 20.2   A British Sopwith Camel World War I biplane fighter. Cellulose acetate lacquers were applied to the fabric-covered fuselage of aircraft in the war (author's own, 2012, Imperial War Museums, London).

Chile saltpetre, $NaNO_3$, also known as sodium nitrate. This nitrate salt occurs in abundance in mineral deposits in Chile and Peru. Before the war, it was imported in large quantities by Britain, Germany, the United States, and other countries for use as a fertiliser and to make nitric acid.

Chloride of lime, $Ca(OCl)_2$, also known as bleaching powder or calcium hypochlorite, was employed as a disinfectant in the trenches, to sterilise drinking water, and in antiseptic preparations such as eusol. It was prepared by the reaction of chlorine with slaked lime (calcium hydroxide, $Ca(OH)_2$):

$$2Cl_2 + 2Ca(OH)_2 \rightarrow Ca(OCl)_2 + CaCl_2 + 2H_2O$$

Slaked lime was prepared by slaking lime with water:

$$CaO + H_2O \rightarrow Ca(OH)_2$$

Chlorine, Cl_2. The yellowish-green gaseous chemical element was used to sterilise water and manufacture chemicals such as chloroform and phosgene. Chlorine gas was also used as a lethal asphyxiating chemical warfare agent. Such asphyxiants were also known as choking agents or lung injurants. The element was manufactured by the electrolysis of brine (a solution of sodium chloride salt, NaCl, in water).

Chloroform, $CHCl_3$. A colourless volatile liquid used by surgeons as an inhalation anaesthetic while carrying out amputations and other operations on wounded soldiers.

Chloropicrin, CCl_3NO_2, also known as trichloronitromethane, is an oily liquid. It releases a powerful vapour that is both a lachrymatory agent (a tear gas) and a lethal asphyxiant. It was deployed by the Germans and Allies, either by itself or with other toxic chemicals, in trench mortar bombs and artillery shells, and for cloud gas operations using cylinders.

Collodion is nitrocellulose dissolved in a mixture of ether and ethanol. The nitrocellulose is not so highly nitrated as guncotton. Collodion was used to make the smokeless propellants ballistite and Cordite RDB, and also in early forms of photography to coat glass plates.

Cordite. A British smokeless propellant consisting of nitroglycerine, nitrocellulose, and petroleum jelly. Guncotton was used as the nitrocellulose in early versions of cordite (Cordite

Mark I and Cordite MD). The later version, Cordite RDB, used collodion instead of guncotton.

Dakin's solution. A non-irritant antiseptic consisting of a dilute solution of sodium hypochlorite, NaOCl, and boric acid. It was used in the treatment of infected wounds during the war. The solution is unstable and so had to be freshly prepared before use.

Diphenylchlorarsine. A chemical warfare agent known as a sternutator introduced by the Germans mid–1917 in their so-called "Blue Cross" shells.

The powder irritates nasal passages causing sneezing, vomiting, and headaches. The Germans fired some ten million or more such shells in the latter part of the war.

Diphosgene. An oily liquid that gives off a lethal asphyxiating vapour.

The toxic liquid was first used by the Germans in their "Green Cross" shells in May 1916 at the Battle of Verdun. The liquid was easy to pour into shells and became the principal killing gas used in shells during the war.

Dynamite. A blasting explosive, patented by Alfred Nobel in 1867, that contains about 75% nitroglycerine and 25% kieselguhr (an absorbent material added to stabilise the explosive so that it can be safely packed, stored, and transported). It has been extensively used in mining and quarrying, and for the demolition of buildings. In the First World War, the British Royal Navy used the explosive during the raid on the German-held Belgian port of Zeebrugge on the night of 22/23 April 1918.

Ether, C_2H_5-O-C_2H_5, more accurately known as diethyl ether, was one of the inhalation anaesthetics employed by surgeons in the war. It was sometimes used in combination with chloroform.

Eusol is an acronym for Edinburgh University Solution of Lime. The antiseptic solution was used by the British in the war. It was prepared by dissolving "eupad" (a powder consisting of equal weights of chloride of lime and boric acid) in water. When the powder dissolved, the chloride of lime released hypochlorous acid, HOCl, a powerful antiseptic compound.

Guncotton is a highly nitrated form of nitrocellulose. The explosive is an extremely flammable yellowish-white compound consisting of filaments that look like raw cotton. It is prepared by nitrating cotton with a mixture of concentrated nitric and sulfuric acids at low temperature. In the war, the French used the explosive to make their smokeless propellant *Poudre B* and the British to make cordite.

Gunmetal is a copper–tin alloy typically consisting of 88% copper, 8% tin and 4% zinc. The metal was used to make guns before the development of low-cost steels in the 19th century. In World War I, the alloy was used to fabricate various components of shells and fuses.

Gunpowder, also known as black powder, was employed as a general purpose low explosive in the war (Figure 20.3). The composition of the powder varied, but typically consisted of a mixture of 75% potassium nitrate, KNO_3, 15% charcoal, and 10% sulfur. It was used as a propellant in firearm cartridges, by the artillery to fire shells, and as an explosive in shrapnel shells. It was also used as a blasting explosive in mining operations. Low explosive powders require ignition to explode. The powder then burns rapidly, but at a rate of combustion that is less than the speed of sound. The nitrate in gunpowder is an oxidant that provides the oxygen for the combustion of the fuels in the mixture: charcoal and sulfur. The combustion of gunpowder produces a lot of smoke. Smokeless powders such as cordite produce far less smoke when ignited.

Hypo, $Na_2S_2O_3$, also known as sodium thiosulfate. The name "hypo" derived from one of the compound's old-fashioned chemical names: "hyposulphite of soda." The compound dissolves in water to form an alkaline solution. During the early chlorine gas attacks in April 1915, troops made primitive gas masks from cloths, articles of clothing, or pads of cotton wool

Figure 20.3 Shrapnell shell on display at the Ulster Tower Memorial, Thiepval, northern France. A tin-plated iron cup at the base of the shell contained gunpowder. When it exploded, the head of the shell blew off and the shrapnel balls were ejected at high velocity (author's own, 2010).

soaked in the solution. The solution was also used as a fixing agent to process photographs during the war.

Lewisite, Cl-CH=CH-AsCl$_2$, also known as chlorvinyldi-chlorarsine or the "dew of death." The oily colourless to light amber liquid was developed as a chemical warfare agent by the Americans in 1918 after they entered the war. The highly toxic liquid is a vesicant (blister agent) and its vapour a powerful lung irritant. It was prepared by the reaction of acetylene, CH≡CH, and arsenic trichloride, AsCl$_3$, in the presence of an aluminium chloride, AlCl$_3$, catalyst. The United States shipped a consignment of about 150 tons of the chemical to Europe in November 1918, but the war ended before it could be used.

Lyddite is named after Lydd, the town in Kent, England, where it was manufactured. The compound is also known as picric acid or 2,4,6-trinitrophenol, and by the French as "melinite."

The high explosive yellow compound was synthesised by nitrating carbolic acid (phenol) using a mixture of concentrated nitric and sulfuric acids. The British and French used it in the war to fill high explosive shells. High explosives, unlike low explosives such as cordite and gunpowder, explode on detonation virtually instantaneously and the shock wave travels through the material at speeds faster than the speed of sound. Picric acid dissolved in an aqueous solution of ethanol was employed as an antiseptic to clean wounds and prevent infections during the war.

Mercury fulminate, $Hg(CNO)_2$, also known as mercury cyanate, is a primary explosive that detonates spontaneously and violently by shock, friction or when heated or hit by a spark. The grey crystalline compound is prepared by treating a solution of mercury, Hg, in nitric acid with ethanol, C_2H_5OH. The compound was carefully mixed with other chemicals to create detonating compositions for fuses, grenades, bombs, and percussion caps in firearm ammunition and cartridges for artillery shells.

Morphia, known as morphine nowadays, is a bitter white crystalline compound that has very low solubility in water.

Water-soluble morphine salts were the painkillers of choice in the war. They were administered by mouth or by injection in various analgesic preparations. Morphine and the related compound codeine are two of the eighty or so alkaloids that occur naturally in the opium poppy (Figure 20.4). Alkaloids are nitrogen-containing organic compounds with complicated chemical structures found in plants. Omnopon, a mixture containing hydrochloride salts of morphine, codeine (a less potent analgesic than morphine), and other opium alkaloids, was one of the water-soluble analgesics used in the war. Opium (the dried latex of the opium poppy) and the morphine derivative diacetylmorphine (commonly known as diamorphine or heroin) were also used as painkillers and sedatives in the war.

Mustard gas, $(Cl\text{-}CH_2CH_2)_2S$, also known as sulfur mustard or dichloroethyl sulfide. In 1917 and 1918, the Germans, French, and British manufactured large quantities of the liquid for use as a blister agent in artillery shells. The Allies prepared the chemical by the reaction of sulfur monochloride, S_2Cl_2, and ethene $(C_2H_4,$ also known as ethylene):

$$S_2Cl_2 + 2C_2H_4 \rightarrow (Cl\text{-}CH_2CH_2)_2S + S$$

Figure 20.4 Opium poppies. The alkaloids morphine and codeine extracted from the plant were commonly used as painkillers in the war (author's own, 2007).

Nitric acid, HNO_3. One of the acids that underpinned the war effort on all sides. It was of critical importance for the production of nitro-explosives such as amatol, ammonal, Lyddite, and TNT. It was manufactured by the reaction of Chile saltpetre (sodium nitrate) with concentrated sulfuric acid:

$$2NaNO_3 + H_2SO_4 \rightarrow 2HNO_3 + Na_2SO_4$$

The Germans also produced nitric acid by the oxidation of ammonia. The Ostwald process, as it is known, combines ammonia with atmospheric oxygen in the presence of a platinum catalyst to form nitrogen monoxide, NO, a gas which is then further oxidised to yield the related gas nitrogen dioxide, NO_2. The nitrogen dioxide is dissolved in water to produce the nitric acid and more nitrogen monoxide which is then recycled. The three steps starting with ammonia are:

$$4NH_3 + 5O_2 \rightarrow 4NO + 6H_2O$$
$$2NO + O_2 \rightarrow 2NO_2$$
$$3NO_2 + H_2O \rightarrow 2HNO_3 + NO$$

Nitroglycerine is an unstable explosive pale yellow oily liquid, prepared by nitrating glycerine at low temperature:

The compound was employed in the war as a component of explosives such as ballistite, cordite, and dynamite. Also known as glyceryl trinitrate, the compound dilates blood vessels and is therefore used to prevent and treat heart diseases such as angina pectoris.

Novocaine, also known as procaine, was one of the most commonly used local anaesthetics in the war. It was administered by injection.

Phosgene, $COCl_2$, also known as carbonyl chloride, is a highly toxic colourless choking gas that was used not only as a chemical warfare agent but also in industry to manufacture dyes and other synthetic organic chemicals. It is produced by the reaction of carbon monoxide, CO, with chlorine, Cl_2. The Germans called the gas "D-Stoff" and first used it mixed with chlorine in a cloud gas operation against the British in December 1915. The French first employed phosgene in artillery shells in February 1916 during the Battle of Verdun and the British used it at the Battle of the Somme later in the year.

Phosphorus, P, is a chemical element that commonly occurs in two forms: white phosphorus and red phosphorus. White phosphorus is a flammable waxy white solid. It does not dissolve in water, but ignites spontaneously in air, combining with oxygen to form phosphorus pentoxide, empirical formula P_2O_5, a white solid that reacts with moisture in air to form phosphoric acid, (H_3PO_4). White phosphorus was used in the war as an obscurant. Smoke shells filled with the obscurant were fired to create smoke blankets over the enemy. The obscurant was also used to create smoke screens to conceal manoeuvres from the enemy. White phosphorus was employed by all sides as an incendiary material in incendiary bombs, shells, and bullets. Phosphorus is extracted from phosphate minerals such as hydroxyapatite, $Ca_5(PO_4)_3(OH)$.

Prussic acid, HCN, also known as hydrogen cyanide or hydrocyanic acid, is a highly toxic colourless liquid that smells like almonds. The chemical is a type of chemical warfare agent known as a blood agent as it prevents blood from carrying oxygen. French "Vincennite" gas shells were filled with a mixture of the blood agent and other toxic chemicals. Over the course of the war, the French used over 4000 tons of the blood agent and fired some four million Vincennite shells. Germans also used the gas to tackle insect infestations in enclosed spaces during the war.

SK, $C_2H_5CO_2CH_2I$, or ethyl iodoacetate, is a colourless oily liquid named after South Kensington, the location of Imperial College, London, where 50 or so chemicals were tested as potential lachrymatory agents for use in the war. The British selected the compound because the iodine required to make it was readily available in the country and the liquid did not corrode the metal canisters that stored the liquid.

Steel is an alloy of iron and carbon. Mild carbon steels contain around 0.2% carbon. They are malleable and ductile and were used for making barbed wire in the war. High-carbon steels contain up to 1.5% carbon. They are harder and were used to make knives and tools. Alloy steels contain not only iron and carbon but also other elements. Stainless steel, for example, contains at least 10% chromium. Nickel steels were used to make armour for tanks and battleships. Manganese steels are hard wearing and were used to make railway lines.

Sulfuric acid, H_2SO_4. In 1843, German chemist Justus von Liebig observed that the commercial prosperity of a country could be judged by the amount of sulfuric acid it consumed. The acid was used before and during the Great War for a multitude of applications, including the manufacture of dyestuffs and fertilisers such as ammonium sulfate, $(NH_4)_2SO_4$. It was used in nitration processes for the manufacture of explosives like TNT and Lyddite, and for the production of nitric acid from Chile

Figure 20.5   Helmets made of alloy steels dramatically reduced the incidence of head wounds in the war (author's own, 2012, Imperial War Museums, London).

saltpetre. The contact process was widely employed during the war to make the acid. Naturally-occurring sulfur or the sulfur in sulfide minerals such as iron pyrite, FeS_2, was first converted into sulfur dioxide, SO_2. The gas was then further oxidised, using a platinum catalyst, to yield sulfur trioxide, SO_3, which was finally converted into H_2SO_4 by the addition of water.

Tetryl is a yellow crystalline compound.

The British used the high explosive in the war as a detonating material and as a booster explosive in some types of high ex- plosive shell and in fuse magazines. Like other nitro-explosives, it required nitric and sulfuric acids for its manufacture.

Tincture of iodine is a weakly antiseptic solution that was used in the war. It was applied to the skin to prevent or treat infections in wounds. The tincture is prepared by dissolving iodine, I_2, in a solution containing potassium iodide, KI, ethanol, C_2H_5OH, and water. Sodium iodide, NaI, is sometimes used in place of potassium iodide.

TNT (trinitrotoluene). The high explosive is a yellow crystalline compound that was first prepared by German chemist Julius Wilbrand in 1863.

The material was used in the war, either by itself or as a component of explosive mixtures, as a blasting explosive in tunnelling operations, and as a bursting charge in high-explosive artillery shells. It was safer and easier to handle, store and transport than Lyddite.

T-Stoff was a mixture of two lachrymators (tear gases): benzyl bromide (left) and xylyl bromide (right).

The German Army fired shells containing the eye irritant mixture, known as "T shells," at Russian troops on the Eastern Front in January 1915. The weather was too cold for the mixture to evaporate sufficiently and so the shells proved ineffective. Even so, the shells remained in use on both the Eastern and Western fronts until 1917.

Washing soda, Na_2CO_3, a white solid also known as sodium carbonate, dissolves in water to form an alkaline solution. In April 1915, troops used cloths, clothing, or pads of cotton wool soaked in the solution for protection against chlorine gas. If washing soda were not available, a solution of sodium bicarbonate, $NaHCO_3$, (also known as sodium hydrogencarbonate or baking soda), or hypo solution was used instead.

Bibliography

K. Adie, *Fighting on the Home Front: The Legacy of Women in World War One*, Hodder & Stoughton, London, 2013.

F. Aftalion, *A History of the International Chemical Industry; from the "Early Days" to 2000*, Chemical Heritage Press, Philadelphia, 2001.

T. Ashworth, *Trench Warfare 1914-1918: The Live and Let Live System*, Pan, London, 2000.

J. M. Barry, *The Great Influenza: The Story of the Deadliest Pandemic in History*, Penguin, London, 2004.

Nineteenth Century Germany: Politics, Culture and Society 1780–1918, ed. J. Breuilly, Arnold, London, 2001.

G. Bridger, *The Great War Handbook*, Pen & Sword, Barnsley, 2009.

G. I. Brown, *Explosives: History with a Bang*, The History Press, Stroud, 2010.

K. Brown, *Fighting Fit: Health, Medicine and War in the Twentieth Century*, The History Press, Stroud, 2008.

M. Brown, *Tommy Goes to War*, The History Press, Stroud, 2009.

E. M. Chamot, *The Microscopy of Small Arms Primers*, Cornell Publications, Ithaca, New York, 1922.

D. Charles, *Between Genius and Genocide: the Tragedy of Fritz Haber, Father of Chemical Warfare*, Pimlico, London, 2006.

The Chemists' War, 1914-1918
By Michael Freemantle
© Freemantle, 2015
Published by the Royal Society of Chemistry, www.rsc.org

P. Coffey, *Cathedrals of Science: The Personalities and Rivalries That Made Modern Chemistry*, Oxford University Press, Oxford, 2008.

G. Corrigan, *Mud, Blood and Poppycock*, Cassell, London, 2003.

E. Croddy with C. Perez-Armendariz and J. Hart, *Chemical and Biological Warfare: A Comprehensive Survey for the Concerned Citizen*, Copernicus Books, New York, 2002.

T. L. Davis, *The Chemistry of Powder and Explosives*, Vol. II, John Wiley & Sons, New York, 1943.

W. Davis, *Into the Silence: The Great War, Mallory, and the Conquest of Everest*, Bodley Head, London, 2011.

J. Ellis and M. Cox, *The World War I Databook*, Aurum Press, London, 2001.

R. Evans, *Gassed*, House of Stratus, London, 2000.

A. Farmer, *The American Civil War: Causes, Course and Consequences 1803–77*, 4th edn, Hodder Education, London, 2006.

S. Faulks, *Birdsong*, Vintage, London, 1994.

R. Fennell, *History of IUPAC 1919–1987*, Blackwell Science, Oxford, 1994.

M. Freemantle, *Gas! GAS! Quick, boys! How Chemistry Changed the First World War*. The History Press, Stroud, 2012.

S. Garfield, *Mauve: How One Man Invented a Colour that Changed the World*, W. W. Norton, New York, 2001.

P. Gatrell, *Russia's First World War: A Social and Economic History*, Pearson Education, Harlow, 2005.

L. F. Haber, *The Poisonous Cloud, Chemical Warfare in the First World War*, Clarendon Press, Oxford, 1986.

R. Harris and J. Paxman, *A Higher Form of Killing: The Secret History of Chemical and Biological Warfare*, Arrow, London, 2002.

G. Hartcup, *The War of Invention: Scientific Developments, 1914–18*, Brassey's Defence Publishers, London, 1988.

M. Hastings, *Catastrophe: Europe Goes to War 1914*, William Collins, London, 2013.

H. Hesketh-Prichard, *Sniping in France: With Notes on the Scientific Training of Scouts, Observers, and Snipers*, Hutchinson, London, 1920.

D. R. Higgins, *Mark IV vs A7V: Villers-Bretonneux 1918*, Osprey, Oxford, 2012.

I. V. Hogg and L. F. Thurston, *British Artillery Weapons & Ammunition 1914–1918*, Ian Allan, Shepperton, 1972.

G. B. Infield, *Disaster at Bari*, Robert Hale & Company, London, 1971.

G. Jukes, *The First World War: The Eastern Front 1914–1918*, Osprey, Oxford, 2002.

The Making of the Chemist: The Social History of Chemistry in Europe 1789–1914), ed. D. Knight and H. Kragh, Cambridge University Press, Cambridge, 1998.

J. LeConte, *Instructions for the Manufacture of Saltpetre*, Charles P. Pelham, State Printer, Columbia, SC, 1862.

J. Lee, *The Gas Attacks Ypres 1915*, Pen & Sword, Barnsley, 2009.

V. Lefebure, *The Riddle of the Rhine, Chemical Warfare in the First World War*, Collins' Clear-Type Press, London and Glasgow, 1921.

The Essential Chaim Weizmann: The Man, the Statesman, the Scientist, ed. B. Litvinoff, Weidenfeld and Nicolson, London, 1982.

G. Lunge, *Coal-tar and Ammonia*, Part I, 4th edn, Gurney and Jackson, London, 1909.

G. Lunge, *Coal-tar and Ammonia*, Part III, 5th edn, Van Nostrand, New York, 1916.

Frontline and Factory: Comparative Perspectives on the Chemical Industry at War, 1914–1924, ed. R. MacLeod and J. A. Johnson, Springer, Dordrecht, 2006.

A. Marwick *The Deluge: British Society and the First World War* (Student Edition), Macmillan Press, London and Basingstoke, 1973.

Y. McEwen, *"It's a Long Way to Tipperary" – British and Irish Nurses in the Great War*, Cualann Press, Dunfermline, 2006.

M. Morpurgo, *War Horse*, Egmont, London, 2007.

A. Palazzo, *Seeking Victory on the Western Front: The British Army & Chemical Warfare in World War I*, University of Nebraska Press, Lincoln and London, 2000.

J. R. Partington, *The Alkali Industry*, 2nd edn, Van Nostrand, New York, 1925.

M. Pegler, *British Tommy: 1914–1918*, Osprey, Oxford, 1996.

R. B. Pilcher, *What Industry Owes to Chemical Science*, The Scientific Book Club, London, 1947.

A. M. Prentiss, *Chemicals in War: A Treatise on Chemical Warfare*, McGraw-Hill, New York, 1937.

A. Rance, *Southampton. An Illustrated History*, Milestone, Hordean, 1986.

M. Rayner-Canham and G. Rayner-Canham, *Chemistry Was Their Life: Pioneer British Women Chemists, 1880–1949*, Imperial College Press, London, 2008.

B. G. Reuben and M. L. Burstall, *The Chemical Economy: A Guide to the Technology and Economics of the Chemical Industry*, Longman, 1973.

M. S. Russell, *The Chemistry of Fireworks*, RSC Publishing, Cambridge, 2000.

C. A. Russell and J. A. Hudson *Early Railway Chemistry and its Legacy*, RSC Publishing, Cambridge, 2012.

G. Sheffield, *The Chief: Douglas Haig and the British Army*, Aurum Press, London, 2011.

War on the Western Front, ed. G. Sheffield, Osprey, Oxford, 2007.

N. R. Storey and M. Housego, *Women in the First World War*, Shire Publications, Oxford, 2010.

The War Poets, (poems of the First and Second World Wars selected by Michael Wylie), Pitkin Publishing, Andover, 1992.

R. W. Thomas and F. Farago, *Industrial Chemistry*, Heinemann, London, 1973.

W. A. Tilden, *Chemical Discovery and Invention in the Twentieth Century*, 3rd edn, George Routledge and Sons, London, 1919.

J. N. Tønnessen and A. O. Johnsen, *The History of Modern Whaling*. C. Hurst & Company, London, and Australian National University Press, Canberra, 1982.

Determinants in the Evolution of the European Chemical Industry, 1900–1939, ed. A. S. Travis, H. G. Schröter, E. Homburg and P. J. T. Morris,Kluwer Academic Publishers, Dordrecht, The Netherlands, 1998.

War Office, *Treatise on Ammunition*, 10th edn, The Naval & Military Press, Uckfield, East Sussex, and The Imperial War Museum, London, 1915.

C. Weizmann, *Trial and Error: The Autobiography of Chaim Weizmann*, Hamish Hamilton, London, 1949.

Chemistry at Oxford: A History from 1600 to 2005, ed. R. J. P. Williams, J. S. Rowlinson and A. Chapman, Royal Society of Chemistry, Cambridge, 2009.

C. Williams-Ellis and A. Williams-Ellis, *The Tank Corps*, George H. Doran Company, New York, 1919.

D. Winter, *Death's Men*, Penguin Books, London, 1979.

S. Young, *Distillation Principles and Processes*, Macmillan, London, 1922.

Subject Index

References to figures are given in *italic* type.